How Flavor Works

T0201347

How Flavor Works

The Science of Taste and Aroma

Nak-Eon Choi
Director, Sias Co. Ltd
Oksan-myun, Heungdeok-gu, Cheongju-si, Chungcheongbukdo,
South Korea

Jung H. Han
Adjunct Associate Professor in the Department of Food Science and Human
Nutrition, University of Illinois at Urbana/Champaign, USA

WILEY Blackwell

This edition first published 2015 © 2015 by John Wiley & Sons, Ltd

Registered office: John Wiley & Sons, Ltd, The Atrium, Southern Gate, Chichester, West Sussex, PO19 8SQ, UK

Editorial offices: 9600 Garsington Road, Oxford, OX4 2DQ, UK
The Atrium, Southern Gate, Chichester, West Sussex, PO19 8SQ, UK
111 River Street, Hoboken, NJ 07030-5774, USA

For details of our global editorial offices, for customer services and for information about how to apply for permission to reuse the copyright material in this book please see our website at www.wiley.com/wiley-blackwell.

The right of the author to be identified as the author of this work has been asserted in accordance with the UK Copyright, Designs and Patents Act 1988.

Library of Congress Cataloging-in-Publication Data

Choi, Nak-Eon, 1965-
  How flavor works : the science of taste and aroma / Nak-Eon Choi, Jung H. Han.
    pages cm
  Includes bibliographical references and index.
  ISBN 978-1-118-86547-7 (pbk.)
1. Taste. 2. Smell. I. Han, Jung H., 1964- II. Title.
  QP456.C477 2015
  612.8′7–dc23

                                                                        2014031608

A catalogue record for this book is available from the British Library.

Wiley also publishes its books in a variety of electronic formats. Some content that appears in print may not be available in electronic books.

Cover image: artichokes at market © laughingmango/iStockphoto; Woman eating salad © aldomurillo/iStockphoto; Eating sandwich © mediaphotos/iStockphoto; and indian spices collection © bonchan/iStockphoto
Cover design by Translation by DIYPIA, Inc., S. Korea

Typeset in 10/13pt Palatino by Laserwords Private Limited, Chennai, India
Printed and bound in Malaysia by Vivar Printing Sdn Bhd

1   2015

Contents

Preface

Nowadays, technological developments progress at amazing rates. High-specification devices mean that people are connected wirelessly 24 hours a day everywhere with various services providing high-speed information. Many ordinary people now struggle to assimilate the large quantity of information that is available to them. There is so much accessible data and information around us, and, consequently there is a lot of bunk science which is not supported by subject experts or by any scientific evidence. In the high-speed Internet and media worlds, the voices of non-experts are generally louder than the true guidance of experts and junk science often captures the imagination of the public to the detriment of factual information. Food and diet are definitely favorite topics of pseudo-science. Some of the most common non-scientific examples are: MSG is unhealthy; high-fructose corn syrup is bad; natural ingredients have miraculous disease curing effects; microwave ovens kill the good active ingredients in foods; organic foods are more nutritious; and so on. These examples of bunk science spread much faster and are more persuasive to some people than genuine food science, especially in the world of the Internet. *Time Magazine* actually published an article entitled "6 Food Myths Debunked (Alice Park, April 7, 2014. http://time.com/50163/6-food-myths-debunked/)". The six false myths that *Time* wanted to correct are: (1) microwaving foods kills nutrients, (2) the more grains, the better, (3) fat-free salad dressings are healthier, (4) avoid white vegetables, (5) juice cleansers are cleansing, and (6) coffee will only make you thirstier. None of these myths are supported by scientific evidence and they lead public opinion away from the true about foods and human health and culture.

As a Certified Food Scientist, I often feel a responsibility to correct such myths and to campaign for true knowledge with evidence-based scientific research for the general public and readers. Of course, whenever I have a chance to lecture or when I use social networks, I do raise my voice against pseudo food science. During my efforts to fix skewed information, I met a person who had already done a lot more work than me on the same subject. His name was Nak-Eon Choi and I met him through the Internet. I became his Facebook friend and also frequently visited his well-known web site (www.seehint.com), which contains a tremendous amount of information and knowledge on food science and culinology. Just after he published his third book *What is Taste* (2013), in Korean, I contacted him and proposed my plan to publish his book in English. Thankfully, he accepted my proposal and I started to work as his translation editor and a project manager for the publication of the book. High-speed Internet environments spread bunk science efficiently, but also work to create valuable networks with positive motivation and true knowledge campaigning. I think this book is a valuable result example of the construction of a positive network and provides momentum in the direction of true knowledge management.

Special thanks go to Jin Chul Choi, CEO of Sias Co., Ltd (South Korea), who supported the original author, Nak-Eon Choi, so that he could concentrate his efforts on this publication during his regular working hours. Without the support of Jin Chul Choi, the author could not have begun to work on his first version of this book in Korean. I also thank Yong Hoon Lim, CEO of Yemundang (South Korea), who published the Korean version in 2013. I also express my gratitude to DIYPIA, Inc. (South Korea), who worked on the first translation of the original book into English and Christian S. Han (University of Texas at Austin) who did the final corrections on completion of the English version. Also I thank the staff of Wiley in Oxford, for their strong support and patience. However, most of all, my gratitude goes to the original author, Nak-Eon Choi, the sole author of the Korean version, he who remains the primary author of this book. I believe that almost all of the credit for publishing this book should go to him. After reading the book, I hope everyone will begin to trust and appreciate all of the hard work and efforts by qualified food scientists on research and product development to provide better and safer food products for consumers around the world.

Jung H. Han, Ph.D., CFS

About the Authors

Nak-Eon Choi graduated from the Department of Food Science and Technology at Seoul National University with B.Sc. and M.Sc. degrees. He worked as a product developer and flavorist at a confectionary company in Korea, and is currently an R&D Director at Sias Co. Ltd, which is a flavor and sauce manufacturer in Korea. He has published several books in Korean, including *Ineligible Information Ruins Our Body* (2012), *33 Secrets of Food* (2012), *What Is Taste* (2013), *Story of Food Additives* (2013), *Umami and MSG* (2013), *Perception, Illusion, and Hallucination* (2014), and *Flavors of Coffee* (2014). He is also a super blogger managing www.seehint.com, which specializes in food and other related sciences.

Jung H. Han obtained B.Sc. and M.Sc. degrees from Korea University, and a Ph.D. in Food Science from Purdue University. He is a Certified Food Scientist (CFS) and an Adjunct Associate Professor in the Department of Food Science and Human Nutrition at the University of Illinois at Urbana/Champaign, USA. He was an Assistant and Associate Professor in the Department of Food Science at the University of Manitoba before he joined PepsiCo Corporate R&D. He has been an Associate Editor of the *Journal of Food Science* since 2004. He has been the editor of various books, including *Innovations in Food Packaging*, 1st Edition (Elsevier, 2005), *Packaging for Nonthermal Processing of Foods* (Blackwell, 2007), *Modified Atmosphere Packaging for Fresh Cut Fruits and Vegetables* (Wiley-Blackwell, 2011), and *Innovations in Food Packaging*, 2nd Edition (Elsevier, 2013).

1 What is Taste?

> Le Créateur, en obligeant l'homme à manger pour vivre, l'y invite par l'appétit, et l'en récompense par le plaisir. (The Creator, when he obliges man to eat, invites him to do so by appetite, and rewards him by pleasure.)
>
> *Jean Anthelme Brillat-Savarin (Physiologie du goût, 1825)*

Taste is the major influence when deciding which food products to purchase. Globally, regardless of geographical and cultural backgrounds, tastier foods will sell better. For example, strawberries can be found in grocery shops whatever the season or location because they are in high demand. Rarely are any bad-tasting foods popular. The food business worldwide is about 4 trillion USD and the huge size of this market reflects how much taste influences the value of this business. In addition to the food manufacturing business, there are many other different types of food industries. Worldwide, many TV stations broadcast food shows featuring the country's native cuisine. Also, most newspapers and magazines publish food review articles. Recently, as well as these conventional journalism sectors, internet web sites have begun to post various stories on delicious foods and restaurants. However, as people obsess more frequently over tasty foods, some serious side effects have also become more common. The most severe side effect is obesity. Since this is caused by excessive eating habits, solving obesity seems very simple: just "eat less." However, today's obesity issue is worse than ever, regardless of the simple solution. This is because the answer to successful weight control requires a change in lifestyle, and this can be extremely difficult for some people. Another issue is that in today's culture food is not eaten just to eliminate hunger, people also eat when they are not hungry. The indulgent satisfaction that comes from eating overcomes

How Flavor Works: The Science of Taste and Aroma, First Edition.
Nak-Eon Choi and Jung H. Han.
© 2015 John Wiley & Sons, Ltd. Published 2015 by John Wiley & Sons, Ltd.

their desire to stop eating or, in the long term, to change their lifestyle in order to ensure healthy weight control. Now you may start to question what is taste and its relation to obesity, and how do you make your food tastier yet healthier?

All life forms have to eat to survive. In order to obtain the essential energy and resources necessary for their vitality, consumption of food is a critical and major activity for all living organisms. If there was no indulgent satisfaction from our eating habits, food consumption would be a most tedious task that was simply necessary for humans to survive. It would become a chore in order to maintain our vitality, as is taking prescribed medicines. For example, remembering to take prescribed antibiotics twice a day for 10 days is always difficult. Without the delight of eating, ingesting essential nutrients and obtaining our daily energy would be very tedious. Furthermore, it would be extremely hard work to motivate someone to secure adequate amounts of food every single day. However, the emotional pleasure of eating compensates for this hard work. This pleasure never subsides, and is experienced whenever we eat. Unlike other stimulants, eating food is an easy way to receive instant, constant and satisfying gratification. However, when we start to think about a definition for taste, this becomes very difficult, even though it is easy to actually sense it. If we define taste only as a gustatory sense through the biochemistry in our mouth, there are five basic taste elements: sweetness, saltiness, sourness, bitterness, and umami. However, can we describe all the characteristic tastes of thousands of cuisines with the combinations of only these five basic tastes? The answer is no. The characteristic tastes of foods consist of various flavors as well as these five elements. These unique tastes are identified by combinations of thousands of volatile odorants transferred to the nasal cavity through a narrow air-passage from the throat when we eat food. Patients suffering from nasal congestion cannot sense the taste of food as well as normal people can. You may have experienced a slightly better taste when you swallow bitter medications whilst closing your nose with your fingers. Even trace amounts of volatile flavors can change the overall taste of a food, and in the food industry these can affect the profit and loss of their business directly.

The total amount of odorants in a food product is at trace levels. For example, lycopene in tomatoes is only 0.004% but it gives the tomato its famous red color. In most cases, the total amount of odorants is less than 0.01% but this is responsible for the characteristic flavors in the food. Most flowers have less than 0.01% of volatile chemicals in their weight. Sometimes we can obtain a total extract, with both volatile and nonvolatile chemicals, which may reach 0.1% of the total flower composition. However, not all of the components of the volatile extract have the characteristic odorants of the flower. It would be difficult to characterize various fruits by their composition analysis data; however, trace amounts of flavors could classify many fruits into their characteristic groups. An infusion of a small amount of a flavor can make bland fruits taste like completely different fruits.

The human brain allocates only 0.1% of its space to the sensory function of smell, while the area for the visual sensory system takes up 25%. From this statistic, the olfactory sense seems very insignificant, sluggish, and undeveloped. However, in many other animals the olfactory sense is the most developed, sensitive, and dominant sensory system. Did the human sensory system for smell really degenerate? I believe the answer is no. As opposed to those of other animals, the human brain, especially the parts in charge of all sensory systems other than smell, is uniquely advanced, resulting in a relatively small portion of the brain being responsible for the function of smell. Most carnivores have a very sensitive olfactory sense, especially to the odor of prey, compared with that of a human. However, the olfactory sense of a human is not inferior to the sense of other animals when we consider various scent recognition abilities and we understand the connection between this sense to memory, emotions, or other brain functions. The olfactory system of a human is much more important than any other sensory system. Knowing this allows us to correctly understand flavors and intuitively utilize the sensory mechanism.

How can we smell an odorant? Obviously we smell through our nose. The actively functioning area for recognizing the odorant is located in the olfactory epithelium, the top of our nasal cavity, which is only the size of a small coin. This area has more than 400 different types of olfactory cells, indicating that there are far more than 400 genes related to these cells. Compare this with only three types of photoreceptor cells responding to light, one type of taste receptor cell for sweetness, and two types of cells for umami. When we compare the number of olfactory cell types to the small number of receptor cell types for other important sensory systems and also the small number of cell types for other essential metabolisms, the relative number of olfactory cell types is notably large. The entire human body has only 23 000 genes. Therefore, it would be a relatively very high number of genes that are required to develop only one sensory system of smell with 400 types of olfactory cells.

G-protein coupled receptors (GPCR) are found in the olfactory cells. In 1994, Alfred Gilman and Martin Rodbell received the Nobel Prize in Physiology for their discovery of G-proteins and the role of these proteins in signal transduction. Richard Axel and Linda Buck received the Nobel Prize in Physiology for their discoveries of odorant receptors (GPCR) and the organization of the olfactory system in 2004. In 2012, Robert Lefkowitz and Brian Kent Kobilka also received the Nobel Prize in Chemistry for their studies of GPCR. It is very surprising that studies on one sensory mechanism can yield three Nobel Prizes. However, the news of Nobel Prizes related to GPCR studies and olfactory sensory mechanisms has not been as widespread as much other news. Thus, the olfactory sense has become a forgotten physiological topic. However, the mechanism of the olfactory system is the first signal process that living organisms respond to in natural environments. Predators identify the location of their prey using their olfactory signals first, and also recognize edible foods from toxic materials after sniffing them. They do this by remembering which smells relate to which experiences. The system of smell can be developed earlier than any other sensory system.

In the same way that a small amount of hormone can differentiate all body metabolisms, a small amount of odorant can completely dictate the flavors and tastes of foods. However, we enjoy eating foods with flavors without recognizing all of the odorants in them. Of course enjoying the flavors of foods is more important than analyzing all the volatile chemicals and we do not need to actually think about the various odorants. Moreover we have all heard about so many cases of incorrect information on foods that has had no validation by experts on the specific subject matter, which ironically influence our lifestyle by making us more anxious than any evidence-based suggestions from food specialists. Therefore, it is necessary and worth understanding more about odorants and the sense of smell.

All tastes can be classified as one of two extreme results: delightful or disgusting. Since most processed foods are made from high quality ingredients and well-established technologies, they are generally delicious and consequently pleasurable. Some people may say that they have never eaten extremely bad-tasting food, or experienced a displeasing sense. However, the taste is relative. When we are hungry, most foods may seem delicious. Therefore, the opposite meaning of delightful or pleasurable is not disgust. It is satiation that produces a disinterest in eating foods. When we feel satiated, we stop eating. If the feeling of satiation is weakened in a person, they will continue to eat, and it is easy for them to become obese. Understanding the mechanism of satiation is the key to solving the obesity issue.

Four basic tastes, as proposed by Aristotle

Democritus, a fourth century philosopher, hypothesized that the sense of taste was related to the elemental shape of food particles. He thought that the shape of a sweet atom was large and spherical, a sour atom was relatively large but a rough polygon, a salty atom was an isosceles triangle, and a bitter atom was a small, smooth and sphere-like polygon. Plato believed this hypothesis and established his own theory of taste. More specifically, he considered that the characteristic tastes were differentiated as the taste atoms penetrated the capillary vessels in the tongue, and that the vessels were connected to the heart. In his book, *De Anima*, Aristotle indicated that there are four basic tastes: sweetness, sourness, saltiness, and bitterness. For 2000 years his four basic tastes theory was not challenged or questioned. People thought that the function of the taste buds was to sense the four different tastes of foods, and that these taste buds were located on the surface of the tongue. This theory received more support after scientists had discovered the taste buds. Under a microscope, a taste bud cell looks like a keyhole, and, therefore, it would be very credible to imagine the penetration of taste chemicals through this hole, in analogy with the model of a key fitting into a lock. This was believed to be an important model for the gustatory sense.

Early in the twentieth century scientists drew a taste map of the tongue, in which specific regions were assigned as the major areas for sensing each basic taste. The taste map clearly illustrated that the tip of tongue is sensitive to sweetness, both sides are sensitive to sourness, the far back region of the tongue is related to bitterness, while all regions sense saltiness. The sense of taste was treated as a very simple process. Many people have tried various ways to avoid contact of powdered medicine with the far back region of the tongue, according to this taste map, to minimize the bitterness of the medicine. Nevertheless, the conclusions of this taste map were premature, being based on experiments with a small number of research subjects. However, recent studies have verified that humans can identify all taste chemicals from all regions of the tongue. Robert Magolskee investigated the taste map issue and concluded that all regions of the tongue have different types of taste buds and can perceive every taste. He said that the taste map episode is a great example of how difficult it is to eradicate stereotypes, even in scientific fields. For 2000 years, inaccurate results have been cited continuously, without any validation. It also took a very long time for umami to be accepted as a fifth basic taste.

In 1907, Kikunae Ikeda, a Japanese chemist, had a keen interest in the taste of dashi. Dashi is a Japanese consommé made from the dried sea vegetable kelp. Professor Ikeda was curious about the identity of umami, which means delicious taste in Japanese, of his wife's dashi. He could sense the delicious taste of dashi, which is different from any of the four basic tastes. To understand the nature of the mysterious taste of the clear sea vegetable broth, he analyzed enormous amounts of brown algae and sea vegetables and studied many types of cuisine. His studies concluded that there is a fifth basic taste other than the already well known four tastes, which is found in asparagus, tomatoes, cheese, and meat. Nobody paid any attention to his claim, but Ikeda continued his research and finally discovered the mysterious chemical taste of the dashi. He found that the mysterious molecule was glutamic acid.

Glutamic acid is a common amino acid that is a component of proteins. However, when glutamic acid is polymerized to a peptide along with other common amino acids, it loses its taste due to the increased molecular weight. When the tasteless protein is hydrolyzed to release free amino acids through the process of fermentation or cooking, we can taste the glutamic acid. Ikeda concluded "Through this research I found two facts: one is that the sea vegetable dashi contains glutamic acid, and the other is that the glutamic acid is the main source of umami." Ikeda's conclusion was the revolutionary discovery of taste physiology, but this was ignored entirely. Western scientists, who trusted Aristotle's 2000 year old theory without having validated it, evidently did not appreciate the concept of umami. While chefs have been developing improved recipes focusing on glutamic acid rich foods, such as parmesan cheese, tomato sauce, meat broth, dashi, and soy sauce, scientists maintained their beliefs in the four basic tastes theory and taught this theory to the next generations of scientists.

In addition to glutamic acid, other chemical compounds have been identified that exhibit umami. Inosinic acid, a nucleotide seasoning component, was discovered from katsuobushi (dried skipjack tuna flake) in 1913, and guanylic acid was isolated from shiitake mushroom in 1957. However, since 1985 umami has been accepted as the fifth basic taste. Umami has been scientifically proven to be a different taste element to the four basic tastes, as scientists found the umami receptor in the taste buds of mice in 1997 and in the human tongue in 2000. The receptor for umami has a very similar structure to the receptor for glutamic acid, a neurotransmitter in the neurons of the brain. In 2002 a second glutamic acid receptor was found, which is structurally related to the sweetness receptor.

Taste is complex

As far as we know, there are only five basic tastes. However, there are thousands (maybe more than 10 000) different cuisines with unique tastes. There are also various favorite tastes, such as the taste of strawberries or apples. Thus in practice, we can differentiate thousands of characteristic tastes from a variety of foods. Scientifically, these tastes are not the senses as detected by the tongue: they are flavors. Humans can identify only five basic tastes with their tongues. Every so-called taste other than these five basic ones is sensed through an odorant recognized by our noses: these are called flavors. We can recognize and characterize the taste of apples; but chemically it is a combination of sweetness, sourness, and apple flavors. Typically, we do not differentiate these basic tastes and flavors from the actual taste of the apples as it is very difficult to analyze the basic tastes and flavors of foods separately, and this is not necessary to enjoy an apple. Furthermore, the general terminology for taste consists not only of the combination of basic tastes and flavors, but also the includes holistic perceptions related to the consumption of foods. The holistic perception of these factors can include the senses of touch, temperature, pain, gastrointestinal response, nourishment, emotion, environmental, hedonic preference, psychological condition, and cultural heritage, in addition to the basic tastes and flavors. Although there is a difference between the general definition of taste and that of basic taste, there is no adequate word to differentiate between them. In this book, the word taste will be used to express the holistic perception of general taste unless it is specified as a basic taste. The flavors of foods will be discussed more than the tastes. This chapter will begin by explaining taste as opposed to flavors, which will be discussed later in the book.

Most food ingredients are tasteless, odorless, and colorless

The majority of people understand that the tastes and flavors of foods are constructed through a combination of the tastes and flavors of each ingredient. These food ingredients are usually biological materials that are produced organically. The ingredient materials, which are the various tissues of plants and animals, are predominantly biopolymers, such as carbohydrates, proteins, and lipids, and they are typically tasteless, odorless, and colorless. The most common food ingredients are water, carbohydrates (from grains, tubers, and roots), proteins (from meats and legumes), and lipids (from oils and fats). Excluding water and their hydrolysates, all of these are polymers are too big to fit into the taste and flavor receptors. Although there are no evidence-based quantifiable reports, the estimate used in this book is that the proportion of polymeric materials in most common food ingredients is 98% of all edible matter.

Carbohydrates consist of carbon, hydrogen, and oxygen. They exist in nature generally in the form of cellulose and starches, and sometimes as simple sugars. Cellulose and starch are polymers with glucose as a repeating unit. Proteins are polypeptides that are linear chains consisting of from thousands to millions of the 20 types of amino acids. DNA (deoxyribonucleic acid) is a linear polymer made up of 3 billion nucleic acids and RNA (ribonucleic acid) is also a linear polymer of nucleic acids but is shorter than DNA. Lipids, consisting of a glycerol and three fatty acids, are also polymers. Natural fatty acids have an even number of carbons in their chains, because lipid metabolism, which synthesizes or degrades fatty acid chains, utilizes two carbons as the substrate for the enzymatic reactions. The number of carbons in most natural fatty acids ranges from 12 (lauric acid) to 18 (stearic acid). Some fatty acids have unsaturated carbons at positions ω-3, -6, or -9. A water insoluble lipid molecule consists of three fatty acids and a glycerol. Lipids are associated with lipophilic materials. In addition to the carbohydrates, proteins, and lipids, water molecules are bound together through hydrogen bonds, building a cluster with 200–1000 molecules, which functions like a macromolecule. Water molecules do not have any individual mobility, but their clustering groups change through association with other molecules or ions that coexist in the cluster. It is very difficult to actually separate and isolate the clusters from other clusters; however, it is much easier to understand the chemical and physical behavior of water molecules when we treat them as clustered macromolecules. Polymers do not fit into the taste bud receptors either, and, therefore, they are tasteless, odorless, and colorless. This also implies that most common food ingredients, which are polymers, are also tasteless, odorless, and colorless materials. Only small water-soluble chemicals from foods are able to physically access the taste receptors in the taste buds. In conclusion, rather than it being the 98% of the common ingredients of all edible materials, the tens of thousands of tastes, flavors, and colors of food ingredients are mainly characterized by the remaining 2%, which have very low molecular weights.

To begin with, a recognizable odorant requires a specific molecular weight. Only a few molecules are detectable by the olfactory sensory system, and these are generally small with molecular weights below 300, with fewer than 16 carbons in their molecular structures. This indicates that any molecule bigger than two glucose units (total molecular weight of 360) is odorless, because the higher molecular weight chemicals are not volatile, and, consequently, are not detectable by the olfactory cells. It is generally known that the smallest molecule to possess an odor is ammonia, with a molecular weight of 17. All odorants are bigger than an ammonia molecule but have a molecular weight of less than 300. Most carbon-based odorants have 4–16 carbons in their chemical structure, and of these the odorants with 8–10 carbons possess the most elegant fragrance. When the molecular weight of an odorant increases, its volatility and the strength of its odor decrease if the actual quantity of odorant remains the same. However, the molecular structures of the larger molecules may provide a greater potential for binding to a site of the olfactory receptors, which also increases the binding strength of the odorant and the effective time the sense will last. On the other hand, when the molecular weight of the odorant decreases, there are more molecules of the same mass, and the potential for binding to a site of the olfactory receptors decreases. The volatility of the odorants with shorter carbon chains increases, but due to the lower binding strength to the receptors the effective duration of perception decreases. The aromas of the shorter carbon chain odorants are stronger but diminish faster, while the longer chain odorants are more delicate and last longer. Overall, the intermediate molecular weight odorants are the most efficient. Highly hydrophilic (water soluble) small molecules do not have good volatility, while highly lipophilic (oil soluble) chemicals cannot penetrate the mucous layer of the olfactory cells. Therefore, the chemicals that are very soluble in water or in oil are not good odorants, although the majority of odorants are partially hydrophilic as well as partially hydrophobic. However, in most foods, such odorants very rarely exist at the ppm or ppb range. Nevertheless, the amounts present are large enough to be detected by the nose. This is the reason why we can enjoy strong odors from vegetables, fruits, and other foods.

The gustatory sensory system is less reliant on the sensing of the properties of a stimulant than the olfactory sensory system is. The taste chemicals from foods should be hydrophilic with high water solubility and a molecular weight of less than 20 000. Table 1.1 shows the general properties of taste chemicals and odorants. In fact most chemicals that provide the taste to food have molecular weights of much less

Table 1.1 General Properties of Taste Chemicals and Odorants

Parameter		Taste Chemicals	Odorants
Physical properties	Boiling point	High Low volatility	Low High volatility
	Water solubility	Soluble	Insoluble
	Polarity	Large	Small
Chemical properties	Molecular weight	Small–medium (1–20 000)	Small (17–300)
	Molecular structure	Simple–complex	Simple
Amount in nature		Large (%–ppm)	Few (ppm–ppb)
Sensorial description		Simple	Complex

than 20 000 as the smaller molecules are easier to detect. For example, glucose (a monosaccharide) is sweeter than maltose (a disaccharide). The oligomer of glucose is not as sweet as glucose due to its higher molecular weight. The glucose polymer loses its sweetness when it has more than 7–10 glucose units in its molecule, and so clearly starch and cellulose, which are polymers of more than 1000 glucose molecules, are not sweet at all. Ionic transfer through ion channels causes the tastes of sourness and saltiness, and, therefore, they are produced by very small ions. The chemicals that provide the umami taste are also small nucleic acids or a single amino acid. In conclusion, most taste chemicals are small with a molecular weight of less than 1000. Also, as many different chemicals can lead to the sense of bitterness, our gustatory sensorial system has been developed to reject bitter materials from our diet, with over 75% of the total taste receptors functioning to sense bitterness. Most food ingredients are not bitter, and even though there are many bitter materials in nature, in reality most natural materials are actually tasteless or bitter.

Variations in odor during fermentation and aging due to changes in molecular weight

The hypothesis for the significance of molecular size with respect to taste and flavor can be easily demonstrated with fermented foods. During the fermentation process, various new substances are produced through the depolymerization of tasteless carbohydrates, proteins, and lipids. These tasteless polymers are hydrolyzed to give small molecules that do have taste and odor. Fermenting microorganisms are also deconstructed to give a variety of small-molecule chemicals. Thus fermentation enhances the taste and flavor of foods.

However, during the fermentation process, very small molecules, in addition to highly reactive chemicals, are also produced. Many types of aldehydes are excellent examples of such small reactive chemicals, which react with other chemicals and become heavier molecules. A typical aldehyde reaction is that of acetaldehyde in aldol reactions, which are significant in the aging of alcoholic products to smooth the flavor of the aged alcohol. The same reactions occur when fragrances and perfumes are stored in chilled environments for 1 or 2 years to enhance their quality. Sometimes over-ripening produces brown pigments and reduces the strength of the odorant chemicals because the chemical reactions increase the molecular weights too much. These reactions, which produce incremental changes to the size of the odorant chemicals, also reduce the astringency.

The astringent tasting chemicals in unripe persimmons and acorns are water-soluble compounds known as tannins. Tannins are phenolic compounds with various side groups, which also include hydrophilic groups in their structures. Low molecular weight tannins are soluble in water and are less astringent but the tannins with molecular weights ranging from 500 to 3000 are very astringent. Although acorns contain a lot of starch, which is a main energy source for wild animals, the astringent taste prevents them from being eaten. Only squirrels, wild bores, and insects can eat acorns because their tongues are adapted to the astringency. To remove the astringency from acorns, ground acorns can be dispersed in water to separate the water-soluble tannins from the precipitated starch. The astringency of persimmons is removed by the opposite procedure of increasing the molecular weight of the tannins. Just like an aging procedure to smooth the flavors of alcohol and perfumes, the method utilizes aldehydes in the persimmons. Through the acetaldehyde reaction, the molecular weights of the tannins increase and the solubility in water decreases resulting in less astringency. Insoluble tannins have a large molecular size and less hygroscopicity (i.e., the ability to attract water molecules); therefore, they irritate the surface of the tongue less and, consequently, reduce the astringency of the persimmons. Persimmons can be used as a natural remedy to relieve hangovers. The reason for the effectiveness of persimmons is that the tannins are efficient at eliminating the acetaldehyde, which is an incomplete byproduct of alcohol metabolism. When persimmons are stored under a controlled atmosphere with a high level of carbon dioxide, the astringency quickly disappears. This is because pyruvic acid does not fully break down during the Krebs cycle (i.e., oxidative energy pathway) due to the lack of oxygen, and it enters the anaerobic fermentation pathway to produce acetaldehyde, which combines with the tannins in the persimmons. Overall, a simple change to the molecular size of food components can have a significant effect, giving us either a pleasant or an undesirable experience.

2% is not a small amount

Previously it was stated that 98% of food products consist of water, carbohydrates, proteins, and fats, which are tasteless, odorless, and colorless, and that a mere 2% is responsible for the taste of the food. However, this statement may be questionable. Generally, beverage products require 10% sugar in order to taste sweet. In addition to sugar, other materials are responsible for the bitter umami, and sour tastes. Thus the total amount of taste chemicals in foods may be fairly significant and be more than 2%. The 2% concept can be better explained by a summation of all the taste chemicals and flavors from the whole body of edible materials. Most food materials are either specific parts of plants and animals (not the whole body), or crops that have been developed and selected for the extremely efficient production of their edible compounds after numerous breeding programs. The actual amount of colorants or flavors in natural products or processed foods is, at most, 0.1% of the whole of the food. The substances responsible for taste are greater compared with the amount of odorants. A sour taste from a food requires 0.1% citric acid, umami needs 0.5% glutamic acid, and saltiness requires 0.9% salt. The number of carbohydrates possessing sweetness is higher relative to other taste chemicals in a food. The reason why there are more sweet chemicals will be discussed later. The important fact in this chapter is that only a small portion of the food product is responsible for the taste, odor, and color.

How can less than 0.1% of the whole food be responsible for the strong and characteristic flavors? How can it be possible that a butterfly can sense the smell of a trace amount of pheromone from several kilometers away? This seems miraculous for many people, but it is actually a basic natural phenomenon, which can be easily understood through a simple calculation. Mercaptan is a typical example of a substance with a very strong odor. If one drop of this chemical is dropped into a gym with the dimensions $30 \times 30 \times 10 \, \text{m} \ (= 9 \times 10^{12} \, \text{cm}^3)$, how many molecules would exist per cubic centimeter of space? The average molecular weight of common fragrant chemicals is 150 and the mass of one drop of mercaptan is approximated to

0.3 g. The number of odorant molecules in one drop is 1.2×10^{21} ($= 6.02 \times 10^{29} \times 0.3 \, g/150 \, g$). Therefore, the number of mercaptan molecules in a cubic centimeter is 1 300 000 000 ($= 1.2 \times 10^{21}/9 \times 10^{12}$). To produce one molecule in 1 cm^3, we would need 1 300 000 000 gyms. This calculation shows that just one drop of an odorant is not a small amount nor is it a small number of molecules either. This explains how a pheromone works. Pheromones are specific substances that can be recognized by a particular animal. If an animal has a certain receptor to the pheromone, it would be perfectly capable of sensing this substance even if it is normally not detectable by other animals. From this example, it is easy to understand how an odorant can be clearly recognized at very low concentrations. Even a concentration ranging from one part per million to one part per billion is sufficient. The same principle applies for both gustatory and visual sensorial systems. A very small amount of bitterness is enough to be recognized and except for our low sensitivity to salty and sweet tastes, our sensory systems can detect very small amounts of stimuli.

There are many animals, such as dogs, that possess a much higher sensitivity to smell than humans (Table 1.2). Thus, it is more accurate to state that they have a less blunted sense to odorants. For example, dogs do not have a more sensitive olfactory sense, they just have more olfactory cells compared with humans. The sensory area in the nose of a dog is 17 times larger than the human olfactory area, and dogs have ten times more olfactory receptors than humans, so it is no wonder that they have a very keen sense of smell and need only tens of thousands of odorant molecules in order to recognize an odorant. In fact, so far, the most sophisticated analytical instruments for detecting volatile chemicals cannot even determine the threshold concentrations that dogs can.

In general, we cannot feel 98% of the natural materials from foods. It is merely the 2% of the food composition responsible for the taste,

Table 1.2 Threshold Number of Odorant Molecules for Olfactory Systems in a Cubic Centimeter of Air

Air–Borne Odorants	Human	Dog	Threshold Ratio (Human/Dog)
Acetic acid	5.0×10^{13}	5.0×10^{5}	1.0×10^{8}
Hexanoic acid	2.0×10^{11}	4.0×10^{4}	5.0×10^{6}
Valeric acid	6.0×10^{10}	3.5×10^{4}	1.7×10^{6}
Butyric acid	7.0×10^{9}	9.0×10^{3}	7.8×10^{5}
Ionone	3.0×10^{8}	1.0×10^{5}	3.0×10^{3}
Ethyl mercaptan	5.0×10^{8}	2.0×10^{5}	2.0×10^{3}

flavor, and color that either impresses or disappoints us. The sensory preference to a food product depends on its flavor components, which constitute less than 0.1% of the food product. Even today, tens of thousands of different cuisines are being developed with new recipes, but the bad tasting foods are simply rejected. However, good taste is just the right harmony between saltiness, umami, sweetness, and sourness. The range of tastes in various foods is dependent on their many types of flavors, but even considering all these varieties, their concentrations are extremely low. Just as butterflies fly several kilometers because of a pheromone, and salmons furiously return to their parent-stream for two weeks because of the smell, we also surrender our sensorial decisions to the small amounts of flavors in our foods.

When we try to control our diet to manage our weight, the temptation of good-tasting foods is greatly increased. We tend to think that sensing tastes and odorants is natural, but behind every sense there is a reason for their development. To maintain a sensorial sensitivity, many more resources and much more effort are required. It is very valuable to think and question: "why do we sense the taste?" and "how do we sense it?" Knowing the answers to these questions can really help in avoiding incorrect misconceptions about foods. There could be an endless repetition of mistakes in the evaluation of the sensory quality of an entire food from the sensory perceptions from just 2% of the food. Sensory perceptions are much like reading someone's facial expression, as saying that you know everything about a person by just through this is nothing more than self-deception or illusion. The more we understand our sensory perceptions, the more we can understand ourselves. From here on, let's dive into the question of why we feel taste.

2 The Origins of Taste: Why do we Taste?

Sweetness is for identifying energy sources (Carbohydrates)

Living organisms require sugars for two main reasons. The first is for the storage of chemical energy. Biochemical energy for cellular activities is obtained from ATP (adenosine triphosphate), which is produced by hydrolysis of sugars. The brain responds with pleasure when we find sugars with our sweet taste receptors. Sensing sweetness indicates that the food will help fill our demand for energy.

The second function of sugars is to provide a structural building block for the body. This is especially important for plants. Cellulose, hemicellulose, and pectin are long-chain sugars that provide the physical strength for the structure of cell walls. The physical strength of sugar polymers is also a very important tool for chefs. They utilize this strength to create desirable textures in food. Animals also metabolize sugars to obtain ATP and use polymeric sugars for their cellular structures. Sugars conjugate with other chemicals that overhang the cell membranes in order to protect the cells.

The human body consists of 55–64% water, less than 1% carbohydrates, 14–18% protein, and at least 2% fat (but the actual amount is closer to 15–30%). So does this mean that we should ingest foods containing mainly fats and proteins? No, because fat and protein are reusable and non-dissipating materials. Therefore, human bodies only need small amounts of proteins and lipids. From the time of their birth to the first birthday, a human will grow the fastest they will ever grow in their lifetime. A newborn baby weighs around 3.2–3.3 kg, and by the time it turns one-year old, the child weighs around 9–10 kg. Thus,

How Flavor Works: The Science of Taste and Aroma, First Edition.
Nak-Eon Choi and Jung H. Han.
© 2015 John Wiley & Sons, Ltd. Published 2015 by John Wiley & Sons, Ltd.

over the timespan of one year the weight of a baby triples. During that time the mother's milk is typically the only food the baby is given and it contains around 7% protein. A baby only needs 7% protein in its diet during this fastest growth period, therefore, it is unreasonable to think that an adult, without any physical development or growth, requires more than 7% protein. If any readers are incredulous of this fact, consider a muscular bull that consumes only grass.

Carbohydrates are required to be the most abundant everyday energy sources in the body. Therefore, sweetness is a pleasurable sense for anyone trying to acquire an energy source. More research has been done and is still being done on sweetness than on any other gustatory sense, because in the food industry the market demand for sweeteners is large. The sensorial principle of sweetness can be explained with the simple model of (HA)- and B-subunits, where H stands for hydrogen and A and B can be other specific atoms (Figure 2.1). Typical A and B atoms are oxygen and nitrogen atoms. When the (HA)-group and B-group in the sweetness receptors interact with the B-groups and (HA)-groups of the sugar, respectively, a signaling reaction is initiated, and that signal is transmitted through the nerves to the brain. Because of this, there are a lot of substances other than sugars that convey sweetness, such as aspartame, saccharine, and other sweeteners. However, due to the three-dimensional geometry requirements for the binding of (HA)- and B-subunits, some mirror

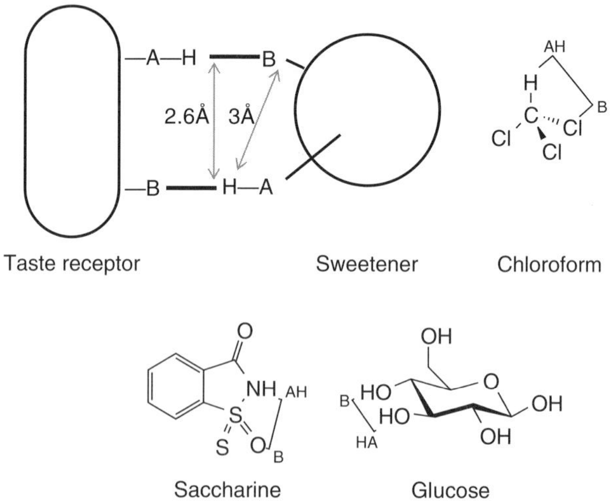

Figure 2.1   **(HA)- and B-subunit model for binding sweeteners to sweetness receptors.**

image isomers of the common sweeteners do not fit into the taste receptors and, consequently, do not possess sweetness.

There have been a considerable number of complications and controversial stories regarding sweeteners throughout history. Honey has been around as long as humans and sucrose was praised as a medicine during the Middle Ages. However, nowadays, sucrose is criticized as being the main cause of obesity and other illnesses. Even fructose, which was enthusiastically welcomed as a form of natural sugar, is now criticized even more than sucrose. To satisfy the consumers' preference for sweetness, without compromising the dependency on sweetness and the high calorie content of sugars, various artificial sweeteners have been developed and introduced to the market. In spite of the low calorific content and scientifically proven safety, artificial sweeteners are constantly viewed as suspicious, unhealthy, man-made chemicals.

Dr. John E. Hayes in the Department of Food Science at Pennsylvania State University has published his research on sweetness perception (Hayes, 2008). He explained that the reason why artificial sweeteners are sweeter than sucrose is because the forces binding to the receptors are stronger than those of sucrose and other natural sweeteners. These stronger binding forces of artificial sweeteners to the receptors last longer than the weaker binding forces of sucrose to the receptors. This is why artificial sweeteners such as saccharine and aspartame score much higher on the sweetness index than sucrose, even for much smaller amounts. Sucrose only binds to the sweetness receptors, which offers a simple and clean sweet taste, but, due to its weak binding forces, the sweetness intensity is lower than artificial sweeteners. Saccharine has very strong binding forces to the sweetness receptors, resulting in a very high score on the sweetness index; however, it also has a binding site to bitter taste receptors and, therefore, provides an undesirable bitter aftertaste. Table 2.1 lists common sweeteners and their molecular weights, and Figure 2.2 shows the chemical structures of three of the most common natural sweeteners.

At this point it is necessary to understand the sweetness index of sweeteners and how the magnitudes are compared. Saccharine is 300 times sweeter than sucrose, but this does not mean that 1/300 of saccharine is needed to create the same taste as sucrose. It actually means that the threshold concentration of dissolved saccharine in water for the initial sweetness sense is 1/300 of the threshold concentration of the sucrose solution. Therefore, saccharine has a threshold concentration that is much lower than sucrose. However, this does not necessarily indicate that the sweetness increases in proportion to

Table 2.1 Molecular Weights of Various Natural and Artificial Sweeteners

Compounds	Range	Mean	Compounds	Range	Mean
D-Fructose	100–180	121	Saccharin	30 000–60 000	–
Sucrose	–	100	Cyclamate	–	3000
D-Glucose	50–82	64	Aspartame	16 000–22 000	–
L-Glucose	–	64	Alitame	–	200 000
D-Galactose	32–67	50	Acesulfame K	–	20 000
α-D-Mannose	32–59	–	Sucralose	40 000–80 000	–
β-D-Mannose	Bitter	–	Thaumatin	200 000–300 000	–
Tagatose	–	92	Glycyrrhizin	5000–10 000	–
Maltose	32–60	43	Stevioside	20 000–30 000	–
Lactose	16–45	33			
Trehalose	–	50			
Xylitol	–	90			
Sorbitol	–	93			

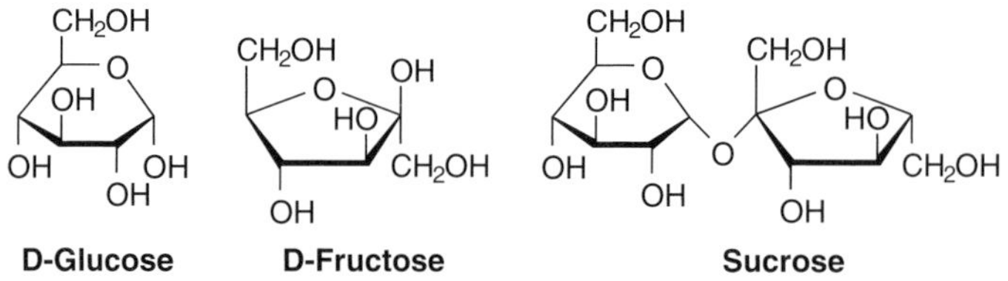

Figure 2.2 Chemical Structures of Sugars.

the concentration. The full and rich sweetness of sucrose cannot be replicated by any concentration of saccharine. On the contrary, once a certain concentration of saccharine is exceeded, the bitterness of saccharine increases and causes an unpleasant experience. Other artificial sweeteners work in the same way. Furthermore, the effects of artificial sweeteners on the sweetness sense are different from animal to animal. For example, mice cannot sense any taste towards aspartame, because their sweet taste receptors cannot bind with aspartame.

Moreover, the level of sweetness and the sensitivity towards sweeteners vary with each person. There are some who have thresholds ten times higher than others. You may have seen someone pour an extensive amount of table sugar into his/her coffee and thought "why does that person like so much sugar?" This situation exemplifies the notion that sensitivity towards sweetness is different for each individual. At room temperature, the average values for the relative sweetness of fructose, sucrose, glucose, galactose, maltose, and lactose are 120, 100, 64, 50, 43, and 33, respectively. Hardly anyone senses glucose as being sweeter than fructose due to the wide difference between the average values. As a side note, the ranges for the relative sweetness of saccharine, aspartame, and stevioside are 30 000–60 000, 16 000–22 000, and 20 000–30 000, respectively.

Different kinds of sugar can provide different types of sweetness. The sweetness of sucrose lasts a long time but it takes some time to bind to the sweetness taste receptors and to initiate the neuro-signal. In comparison, the sweetness of fructose is recognized quickly and at a greater magnitude but the taste also disappears very fast. On the other hand, the sweetness of maltose syrup is sensed slowly and faintly but its sweetness persists for a longer time. The quick reaction of fructose is utilized in many foods to strengthen other flavors that are present especially fruity, salty, and spicy flavors. The sensitivity to tastes and flavors is sharply enhanced because the sweet taste sends signals to the brain to relay that this food is a good source of energy and is worth paying special attention to.

Of all the different types of sugars, sucrose has the most useful properties. Sucrose is the second sweetest sugar after fructose, and unlike other sugars, it creates a pleasant sweetness even in extremely high concentrations as tasted in candy or jam. Sucrose also dissolves very well in water. It can be dissolved in a 2 to 1 ratio of sucrose to water at room temperature. When sucrose is heated without water, it starts to melt around 160 °C, and at about 170 °C it starts to caramelize and forms a distinct taste and odor. When a sucrose solution is heated up with a small amount of acid, it is converted into fructose and glucose. The high viscosity of sucrose in water is a very important property in the art of cooking. When sucrose and other sugars are mixed with the minimum amount of water, they become extremely sticky. For example, spilt sugar water leaves a sticky residue after it has dried up. Interestingly, the reason why diabetes is so frightening is that an excess amount of glucose can make blood sticky and cause difficulties in the circulation of blood. However, in cooking, the water binding capacity of sucrose helps breads and biscuits to maintain their moisture and softness. It acts as an adhering matrix to all ingredients of the food in one bulk material and can make the food shiny and glossy. It also helps with the preservation of food by binding with free water molecules, which spoilage bacteria need to survive. The water binding ability of sucrose also prevents ice cream from freezing too hard. Thus sucrose has a variety of functions other than being a simple sweetener.

Fructose is an isomer of glucose, which means it has the exact same chemical formula as glucose, but it has a different three-dimensional structure. It is the sweetest sugar out of all the common sugars and dissolves best in water. Compared with sucrose and glucose, our bodies metabolize fructose slower. Therefore, it increases the blood sugar level slower than other sugars. This is why fructose is recommended for diabetics, but under controlled consumption. Fructose exists in almost all fruits and becomes sweeter when the fruit is chilled. In fact, the odors of fruits also become stronger when they are chilled. Nonetheless, this does not mean that the concentration of fructose increases with a decrease in temperature. The concentration of sugar within a piece of fruit is constant. What changes is its relative sweetness due to the structural changes of fructose in response to temperature.

As for most sugars, fructose transforms its molecular structure into various arrangements in water. It exists in ring forms (i.e., hexagons or pentagons) and also in a chain form. The binding forces of sugar molecules to the sweet taste receptors of humans vary depending on their structural status. The shape exhibiting the strongest magnitude of sweetness is a six-membered ring (i.e., pyranose) at low temperature. If the temperature rises, the relative sweetness of sugars decreases as the structure of the sugars convert to a five-membered ring (i.e., furanose). Once the temperature reaches 60 °C, the relative sweetness of fructose is nearly halved (Figure 2.3). On the other hand, the sweetness of glucose and sucrose does not change so drastically. At refrigeration temperature, the relative sweetness of fructose is nearly double that of glucose or sucrose. Therefore, in cold drinks, compared with the sweetness of sucrose, half the amount of fructose can exhibit the same magnitude of sweetness as that of sucrose, but with less calories. This is

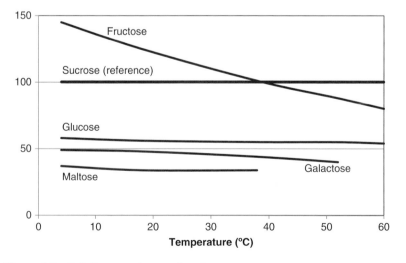

Figure 2.3   Relative sweetness of various sugars.

why high fructose corn syrup is widely used for cold beverages. However, the sweetness of fructose is much lower compared with sucrose in hot drinks; thus, sucrose is commonly used in coffee or tea.

But why do our sweet taste receptors have a weak binding force to sucrose? If the binding force of sugar is 1/10 that of artificial sweeteners, we would require only 1/10 the amount of artificial sweeteners to obtain the same magnitude of sweetness and could reduce our consumption of sucrose to 1/10. This would clearly solve the issue of obesity and other harmful effects of sugar. In order to understand this relatively weak binding force of sucrose to our sweet taste receptors, we should first understand the function of taste in energy ingestion. Bitterness typically represents a warning of toxic components in a food, while sweetness sends information on the nutrition and high-energy values to the brain. It is very important for animals to have simple detection mechanisms for toxins as well as mechanisms for the existence and level of nutrition. Therefore, the strong binding force of toxins to bitter taste receptors enhances the perception and lengthens the retention time of bitterness, even with a very low concentration of toxic stimuli, making it easier to refuse toxic foods. However, the weak binding force of sugars to the sweet taste receptors makes the sweetness disappear at an appropriate retention time, and allows humans to continue to consume the sweet food. For example, if someone ate a grape and its sweet taste stayed in their mouth for a very long time, no one would be able to finish a whole cluster of grapes because they would be too sweet.

We eat carbohydrates more than any other nutrient because it is a main energy source. According to a recommended dietary intake from Japan in 1965, carbohydrates were 72%, proteins were 13%, and lipids were 15% of a person's diet. Before 1965, more carbohydrates were consumed than proteins and lipids. People ate carbohydrates roughly seven times more than proteins or lipids. Humans have always searched for sweets such as honey and grain syrups as well as fruits. If our sensorial sensitivity to sweetness had been maintained at a very precise and very low threshold level, we could not eat the amount of sweet foods we consume today. This situation would make it extremely difficult for someone to take in the required amount of energy through their diet. It is also a part of our evolution of becoming more or less sensitive to certain tastes with respect to the environment.

Starch is a polysaccharide consisting of glucoses in a mixture of linear and branched formations. In the linear portion of the starch polymer fractions, glucose monomers form helical coils through α-1-4 glucosylation, and these linear helices form a bundle of tubes, with several helices, or a sheet. Cellulose is a linear polymer of glucoses (actually a linear polymer of maltose) and is organized as a stacked matrix (i.e., 5×5 stacks of linear polymers), which is called a microfibrile. Although cellulose is a linear chain, it does not form a helix due to the repeating unit of β-1-4 glucosylation. The most abundant organic materials on the earth are cellulose and starch and both are polymers of glucose. Glucose can be converted into many kinds of sugars: glucose can be converted into fructose: fructose can be converted into mannose: and mannose can be converted into galactose. Vitamin C can be synthesized from galactose and is the easiest vitamin to produce and, therefore, the cheapest. In the context of this chapter, it is more important to question why it is only primates who have given up the ability to synthesize vitamin C from glucose and only use glucose as an energy source. Various forms of organic matter are developed from glucose, making it the most common organic compound. In addition, glucose is one of numerous substances with many divergent functions.

Umami is a tool used to search for proteins

Proteins have physiological functions in our body and are essential biochemical substances. Humans must ingest proteins in their diet. Nonetheless, most people cannot recognize the taste of proteins. Proteins are tasteless. This is because proteins are polypeptides made up of several hundred or more amino acids. Any polypeptide with more than ten amino acids does not possess any taste because of its large size, which is too big to bind with the binding site of the amino acid taste receptors. Free amino acids do possess taste, even at very low levels of concentration. Of these amino acids, low water soluble amino acids taste bitter; low molecular weight or highly water soluble amino acids taste sweet. Acidic amino acids, such as glutamic acid and aspartic acid, dissolve well in water, and are the well-known amino acids that produce the umami taste. The glutamic acid molecule is a little longer and has a umami level that is three times stronger than aspartic acid. There may be no substances that can surpass or even match the same intensity of umami as glutamic acid.

All living organisms have proteins in their bodies, and proteins are made up of amino acids. Glutamic acid is the most commonly existing amino acid in natural proteins, and, therefore, is the amino acid that is most produced in most abundance after digestion of proteins (Table 2.2). Glutamic acids can even be produced from carbohydrates

Table 2.2   Composition of Glutamic Acid in Various Proteins

Proteins	Glutamic Acid (%)	Proteins	Glutamic Acid (%)
Beef	15.5	Casein	21.5
Whole milk powder	17.8	Wheat gliadin	43.7
Breast milk	17.0	Wheat glutenin	35.9
Chicken	15.9	Corn zein	31.3
Whole egg	11.9	Corn glutelins	12.9
Egg white	12.7	Rice protein	14.5
Pork	15.5	Soy glycinine	19.5
Mackerel	15.5	Peanut arachin	19.5
Coffee	34.8	–	–
Grapes	14.1	–	–
Tomato	37.0	–	–

or lipids. An average adult weighing 70 kg has around 11.7 kg of protein, and 1.6 kg of those proteins are glutamic acid. The average daily consumption of glutamic acid is around 15 g, but around 50 g of glutamic acid are created from other amino acids within the body. This exemplifies how much our body utilizes glutamic acid. Typically, 10–40% of proteins consist of glutamic acid. Wheat protein has around 40% glutamic acid, and a tomato contains at least 37%. Figure 2.4 shows the chemical structures of aspartic acid, glutamic acid, and MSG.

Compared with other amino acids, glutamic acid is present at an overwhelmingly higher level in our foods. Beef, milk, cheese, yogurt, and every other product derived from cows contain high concentrations of glutamic acid. Similarly, poultry meats and eggs also contain high levels. Fish and other seafoods contain high levels of glutamic acid regardless of whether they are fresh or saltwater fish. Soybean products such as bean sprouts, tofu, soy sauce, soy paste, and soy milk contain high levels of glutamic acid and high concentrations can also be found in various flour based products such as breads and noodles. Coffee, cocoa beans and even fruits are high in glutamic

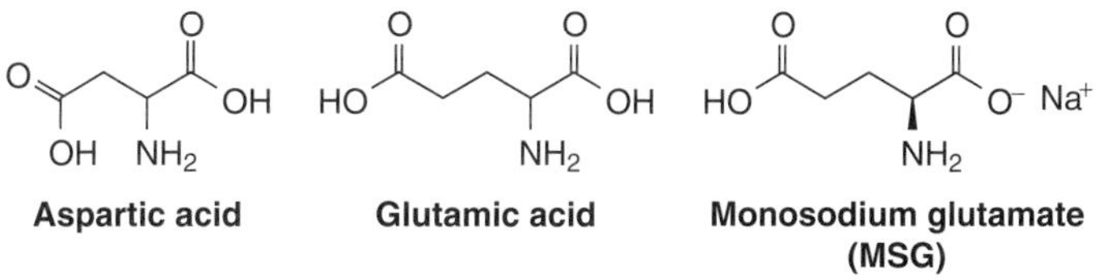

Aspartic acid   **Glutamic acid**   **Monosodium glutamate (MSG)**

Figure 2.4   Chemical structures of aspartic acid, glutamic acid, and MSG.

acid. Breast milk contains less proteins, therefore, less total glutamic acid, than dairy milk, but has ten times more free glutamic acid than dairy milk. Human babies are trained to identify glutamic acid from their mother's milk in order to learn how to identify proteins through umami tasting.

Umami is the exact reason why people enjoy eating meats. Meats are high in protein. When meats are cooked, glutamic acid is produced from their proteins creating their distinct tastes. For example, the concentration of free glutamic acid increases when the meat in fresh pig legs is aged to produce hams. The cheese fermentation process also increases the level of free glutamic acid in the cheeses compared with the young milk curds. Parmesan cheese is a food product with one of the greatest amounts of glutamic acid, with 1.2 g of free glutamic acid for every 100 g. However, the free glutamic acid in Parmesan cheese is only 10% of that of the total amount of glutamic acid in its protein polymers. Tomatoes naturally contain high levels of glutamic acid and the level increases with the maturity of the tomatoes. The taste preference of soy sauces and pastes increases with their level of aging, because the level of free glutamic acid increases during the fermentation process through protein hydrolysis, which is originally present in soy proteins as one of the peptides. The same phenomenon is found with fish sauces and fermented fish. Here the glutamic acid is produced by the hydrolyzation of their proteins during the fermentation process. The reason fish is fermented into pickled fish is to decompose as much glutamic acid as possible from the fish proteins before it is consumed. When making meat broths, various meats such as beef, chicken, and even bones, are boiled in water for a long time to extract more proteins and eventually their glutamic acid. In western culinary arts, the whole basis of sauces is a result of the broth techniques.

In Asian countries, the research and development on monosodium glutamate (MSG), a highly water soluble sodium salt of glutamic acid, is far more advanced than that in Western countries. Since the very early 1900s, Japanese scientists identified MSG as a umami chemical and developed commercial production methods. Most Asians use MSG in their foods to enhance umami. However, in spite of this advancement of MSG research and commercialization in Asia, the commercialization of umami itself started in Europe before MSG was discovered in Japan. Around 1886, hydrolyzed vegetable protein was developed in Europe by Julius Maggi, and in 1908 beef stock was first produced commercially. He developed solid bouillon cubes (stock cubes) so that people who could not afford meats could enjoy a delicious and nutritious soup. This beef stock bouillon cube brought a huge innovation to the food industry because it made cooking more convenient and because the hydrolyzed vegetable proteins taste like meat. Of course these hydrolyzed vegetable protein cubes contained glutamic acid and possessed umami.

In addition to the hydrolysis of proteins, a fermentation method can be used to create a more purified and affordable glutamic acid. Glutamic acid is produced by microbial fermentation, such as in the kimchi or soy sauce production processes. All organic organisms produce glutamic acid through the Krebs cycle. However, the majority of the glutamic acid is secreted from the tissues, and only a limited amount is accumulated in the tissues. Microorganisms secrete surplus amounts of glutamic acid from their cells. When the post-pathway after glutamic acid synthesis is suppressed, the microorganisms continue to synthesize glutamic acid molecules and release them from their cells. This method can make commercial production of glutamic acid affordable and thus it is produced naturally by a fermentation process, without any of the health related issues or controversies related to synthetic chemicals. However, this glutamic acid must then be processed through the chemical reaction of sodium addition in order to resolve its solubility issue.

All substances must be dissolved in water to communicate their taste. However, when glutamic acid is purified, it forms crystals and cannot dissolve in water. To make the glutamic acid crystals soluble and to possess umami, sodium ions are utilized. After the addition of sodium ions to the glutamic acid, the sodium salt of the glutamic acid instantly dissolves in water through the electrostatic repulsion of glutamic acid molecules. Once the glutamic acid sodium salt has dissolved in water, it forms the same natural glutamic acid and free sodium salts. There is no difference between natural glutamic acid and the one in the sodium salt, and it cannot even be distinguished through the most cutting edge analytical technology: the forms are perfectly identical. Trying to distinguish the difference between natural glutamic acid and the one from the sodium salt using any sensory methods is simply nonsense. The sodium salt of natural glutamic acid is called monosodium glutamic acid, which is MSG.

No matter what food is consumed, an average of about 20% of its protein is glutamic acid. Thus, 20% of proteins contain MSG. Compared with the amount of glutamic acid that we ingest from proteins, the amount of glutamic acid from just MSG is an extremely small portion. The glutamic acid in proteins is combined with other amino acids; therefore, it is tasteless and is not sensed by the tongue, and only a very small amount of free glutamic acid that has decomposed from the proteins can transmit umami. That amount is less than 1/200 of the mass of the food. Thus, 1 g of MSG would give a umami taste that is equivalent to the total umami in 200 g of meat. Since only a small amount of MSG is required to provide an adequate amount of the umami taste, the addition of MSG is not enough to make a difference in the total level of glutamic acid in the food. The amount of glutamic acid in the form of MSG necessary to add the umami taste is only about 0.4% of an entire soup. Thus, we would have to consume 10 liters of soup in order to ingest the daily recommended amount of 40 g of MSG. The dispute regarding substantiation of the toxicity of MSG is no more significant than the consideration of the toxicity of proteins.

In addition to glutamic acid, some nucleic acids have umami. The most common umami tasting nucleic acids are 5-inosinic acid (IMP, inosine 5′-monophosphate) and 5-guanylic acid (GMP, guanosine 5′-monophosphate) (Figure 2.5). In order to understand the umami taste in nucleic acids, it is necessary to understand the structure and functions of adenine. Adenine is used in several parts of our body. It is the basic form of ATP that functions as the battery of life. It is also the basic molecule of the essential cAMP (cyclic adenosin monophosphate), the secondary neurotransmitter, and a part of NAD (nicotinamide adenine dinucleotide), FAD (flavin adenine dinucleotide), and coenzyme A. These help the Krebs cycle to function smoothly by acting as co-factors. Furthermore, adenine is one out of the four nucleic acids in charge of our genetic information. The theory that there are taste receptors for such an important nucleic acid is obvious since it is unquestionably important to sense adenine. However, the actual nucleic acids that are tasted are not adenine, but the modified forms of adenine, inosinic acid and guanylic acid. Both of these molecules have the sweetness binding sites of (HA)- and B-groups. Just like MSG, nucleic acid based umami seasonings are also sodium salts of the nucleic acids. However, it is notable that the nucleic

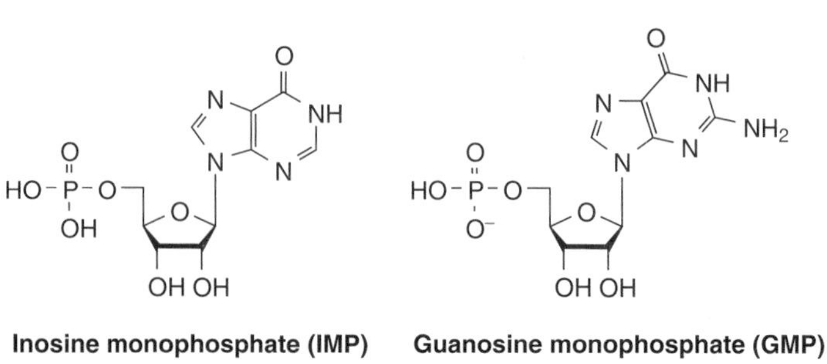

Inosine monophosphate (IMP)   Guanosine monophosphate (GMP)

Figure 2.5   Chemical structures of IMP and GMP.

acid sodium salts have not received any criticisms from nonscientists, even though in their commercial production they undergo the same chemical reaction process as MSG to convert the natural nucleic acids into their sodium salts in order to increase their water solubility.

The intensity of umami increases when amino acid umami (i.e., MSG) and nucleic acid umami (i.e., IMP or GMP) are mixed together. This is why many commercial umami seasonings contain both types. The synergism of umami is commonly utilized not only in the seasoning industry, but also in cooking in general. For example, the addition of dried anchovies and mushrooms into kelp broth will significantly increase the umami, far more than just the umami of the pure kelp broth. In order to understand the mechanism behind this, it is necessary to comprehend the unique structure of umami receptors. Kelp broth contains a lot of MSG, while katsuobushi (Japanese dried tuna flakes) and dried anchovies have lots of IMP and mushrooms have GMP. This is why mushrooms and kelp are used together in recipes. They are especially important in brewing clear soups for very basic broths.

There are some people who think that MSG is a destructive excito-toxin to brain tissues. However, this is a dangerous idea that is not backed up by any understanding of glutamic acid. Glutamic acid is the most abundant neurotransmitter in the nervous systems of vertebrates. At least 50% of the consciousness in human beings is managed by glutamic acid. For mammals, glutamic acid is essential for the cognitive functions that help with the ability to learn and remember. Long-term memory is a result of the functions of the hippocampus, neocortex, and changes in other parts of the brain. Glutamic acid is also known as a precursor of GABA (γ-aminobutylic acid), which is the essential sedative neurotransmitter. GABA is enzymatically synthesized from glutamic acid in a reaction with one simple stage. Our brain functions are regulated by the balance of these two neurotransmitters, stimulating glutamic acid and sedative GABA. In addition, the glutamine that is converted from glutamic acid has an important role in the nervous system. Approximately 1/3 of the glutamine in muscles is released and used by the nervous system during stress, surgery, or illness. This loss of glutamine leads to considerable atrophying of muscle weight. Under such conditions, glutamine supplements are given in order to substantially decrease the amount of glutamine lost in muscles. In other cases, glutamine supplements are used to treat arthritis, immune diseases and disorders, various under-developmental conditions, impotency, schizophrenia, and radiation damage after cancer treatment.

However, most importantly, the glutamic acid (MSG) in foods cannot transfer to our brains. The brain–blood barrier cannot be penetrated by MSG, or many other substances in the blood stream, in order to enter the brain. The brain is self-sufficient and it can produce many necessary materials without the need for these to be supplied from food. The most critical chemical to brain tissue that must be supplied by the blood is glucose. A considerable number of studies regarding Alzheimer's disease have related the disease to the lack of glutamic acid in the brain tissues. However, it is pointless to try to consume more MSG to improve learning capacity or to prevent Alzheimer's disease due to the impermeability of MSG through the brain–blood barrier. It is true that glutamic acid is a stimulating neurotransmitter. The reason we are conscious, thinking, and talking, is due to the function of glutamic acid, therefore, it is a very immature argument to denunciate glutamic acid as a neurotoxin without understanding its pharmacokinetic efficiency.

Carbohydrates are for sweetness, proteins are for umami, but what are lipids for?

Until recently, lipids were thought to be tasteless and only felt during the oral process. However, Professor Pepino at the Washington University School of Medicine revealed that the CD36 gene in the human tongue is required to sense lipids (Pepino *et al.*, 2012). In December 2011 she published her research results on people with a high level of these CD36 genes being able to distinguish a sample with lipids from samples without lipids. The test subjects, who had their noses plugged in a red-lit environment, could distinguish the differences in the taste. The existence of CD36 had already been discovered in rats and cats. If the CD36 was naturally absent in these animals, they would not have this preference for eating high fat foods because they would no longer feel the taste.

However, even with this information, it is still not sufficient to state that we can taste lipids. Even though lipids are tasteless and greasy, they provide a lot of food with aromatic, rich, and smooth characteristics. The free fatty acid receptors (FFARs) in internal organs are connected with the sensory cortex of the brain, and these receptors can sense lipids and other oil-soluble nutrients. In fact, many FFARs were discovered in human intestines and the pancreas and in rodent brains, although only a few FFARs have been found in human tongues. These receptors can sense not only the lipid-based nutrients, but also the quantity, osmotic pressure, temperature, molecular shape, size, and texture of the nutrients. These senses allow our brains to memorize the identity of the food and control the digestion process.

Recollection is commonly thought to be done by the brain, but the body also has its own memory as well as its own non-autonomic nervous systems. If you had a bad experience after eating a specific food, even if you were too young to be aware of your surroundings, you may still not enjoy eating that food as a grown up. You might not know why, but your body (i.e., the non-autonomic nervous systems) remembers that experience. Because the senses in the intestines are connected to the nervous systems, they can initiate various hormone secretions, daily digestive processes, and also satiety. The functions of nutrient receptors in human intestines are very important for controlling digestive systems.

Apart from CD36, some other G-protein receptors have been discovered in the human intestines and pancreas (Covington *et al.*, 2006). GPR40 (i.e., FFAR1) can sense fatty acids with a carbon number of C12–C18, GPR41 (i.e., FFAR3) and GPR43 (i.e., FFAR2) can sense short-chain fatty acids, and GPR84 can sense medium chain (C10–C12) fatty acids. GPR119 and GPR120 (i.e., FFAR4) can sense α-linolenic acid (ALA) and polyunsaturated fatty acids (PUFA). The perception of fatty acids by FFARs is not simply to sense lipid molecules as an energy source, but also as biological materials for other purposes.

Therefore, it would be a big mistake for new product developers to remove fat from their products in the hopes of developing low-calorie products without understanding that the body can sense the type and quantity of fat. Not only does our non-autonomic nervous system remember the taste of foods, but also their calorie content. In addition, even a simple reduction of fat will decrease its flavor intensity and texture. Food researchers have been able to make reduced fat foods with bearable tastes and textures. However, the non-autonomic nervous system does not allow a sudden reduction of fat from one's diet. An extreme cut in carbohydrates such as the Atkins' diet could fool the brain to lose weight, but ultimately the body's desire for carbohydrates will succeed and cause a counteraction, like a yo-yo. Consuming low-fat products might be able to deceive our tongue. However, since they cannot deceive our intestines and non-autonomic nervous system, they will increase the body's cravings to eat more calories and fat.

A little known fact is that each lipid has its own individual odor. Why do foods deep-fried in beef tallow or lard taste and smell better than foods fried in vegetable oil? The answer is that beef and pork fats have their own aromatic compounds that are different from those of vegetable oil. More specifically, there are no odorants in tallow or lard, but their meat flavors are created from the Maillard reaction with sugars (mainly glucose) and cysteine (a sulfuric amino acid). Beef tallow gives off a distinct beef flavor, lard gives off a pork flavor, and chicken fat a chicken flavor. When vegetables are fried in beef tallow, they have a beef flavor. However, when the same food is fried in refined vegetable oil, it has a lower taste preference score because of the lack of beef flavor. Furthermore, lipids have the very efficient ability of preserving odorants. Most odorous chemicals are oil-soluble. The odors that are produced during the cooking process are contained within the lipid and become a part of the whole flavor. This is why meat with more marbling tastes better than meat with less marbling. Lipids affect the texture of fatty foods during various oral processes due to their lubricating effects. In addition to the smoother texture of the food, lipids also increase the whole dynamics of the taste intensity. For example, butter, dressing, and mayonnaise are emulsions. Emulsions consist of a water phase and an oil phase. Both phases are homogeneously mixed macrospectively by surface-active agents that are emulsifiers or emulsifying ingredients. The substances that we naturally enjoy such as salt, sugar, and MSG, dissolve well in their water phase. The water phase has a very low volume, especially in water-in-oil emulsions. If 2.5% of a stick of butter is salt, the salt is not in the butter fat, which takes up nearly 80% of the volume of the stick of butter, but instead the salt is associated with the water phase, which is 20% of the butter. Therefore, the salt concentration of the water phase is 12.5%, which is higher than the amount needed to saturate all receptors. Even though the tongue is stimulated intensely by the 12.5% of salt in the water phase, it does not feel this intensely strong saltiness since the total concentration is still 2.5%. Nonetheless, it may take a longer time to wash out the saturated bound sodium in the taste receptors, which elongates the saltiness profile peak for longer and it is smoother. Of course we like the dynamic conditions of both intensity and smoothness.

Saltiness: the ocean was the source of all life

Salt is an essential source of all biological activity. There is a theory that fish are ancestors of all vertebrates. Approximately 3.5 hundred million years ago, the life forms in the ocean evolved to walk on land.

This may be the reason why certain human body fluids and amniotic fluid are similar to seawater. The salt concentration of these fluids is around 0.9% for humans, while, for seawater, the salt concentration is around 3.5%, because it has gradually been concentrated over time. In other words, the water that takes up more than 65% of our body has the same components as seawater. The intravenous glucose injections that are given in hospitals have 5% glucose in 0.9% saline solution. It is not just blood, but also every tissue and cell that maintains a concentration of 0.9% salt. In fact, even our sweat, tears, and nasal fluids contain salt, and the human fetus grows in salt water inside the womb.

If our bodies had a critically low sodium level, we would die in a couple of minutes. Without sodium, our organs cannot work because there would be no electrical potentials in our nervous systems and tissue membranes. It is very dangerous to supply an excessive amount of water in a short period of time to the body, even after severe dehydration, because the concentration of sodium becomes too low to generate electro-potentials for proper neurotransmission. For plants that do not possess the ability to move, they depend on potassium for electric potential generation because it is easily obtained from the soil. Animals, however, can move so they rely on sodium. This is why they consume salt in any way possible.

Our body needs around 1 g of sodium chloride per day. When most human beings were still in their hunter/gather stages, they could obtain sodium naturally from their carnivorous diet, but ever since we have become settlers and have cultivated crops, it has become very difficult to get the required sodium from our grain-based diet. That is why producers of salt exercised enormous economic and social power in the Eastern and Western regions of the globe. For a long time, salt used to be an extremely important trading item. It had a value equivalent to that of gold and was also utilized as a medium for exchange. It was the symbol of wealth and power. During the twelfth century, salt from southern Morocco was traded for gold in Ghana. A slave was traded for an amount of salt equivalent to the size of the slave's foot. For example, after its insanitary swamps were converted into salt production areas, Venice turned into a wealthy city–state and gained influence over its neighboring nations through its salt production. The commercial transactions of salt and the competition to control salt production advanced the origins of capitalism. Since the sixth century B.C. in Europe and the Warring States Period in China (B.C. 770–221), salt production and trade were monopolized by governments, and, in many countries, its production and supply were regulated. The fact that salt played an important role in the development of civilization can also be identified linguistically. The word salary originates from the time that roman soldiers received salt as their wages. Words such as salad (originally salad was simply seasoned with salt), sauce, salsa, sausage, salami are all derived from the same Latin root, sal, which means salt.

The desire for salt is not limited just to human beings. For all animals, the most urgent desire is to obtain water, quickly followed by salt. The Cingino Dam located in the southwest area of Antrona Schieranco, Italy, has an almost vertical slope with a height of around 50 m. On the slopes of this dam, wild goats, Alpine Ibex, can be seen standing around the massive walls of the dam and licking the dam wall. The reason these goats are huddled together next to the walls of the dam, despite the danger, is none other than the desire to obtain salt from the rocks. Every animal instinctually knows that sodium is critical for sustaining life. Sodium is constantly excreted in urine and sweat and must be replenished. For herbivores, the need for salt is even more critical because it is much harder for them to obtain sodium from their diet. They often place themselves in life-threatening situations just to obtain salt. In nature, hunters lure animals into open places by using lumps of salt or wait for them near places where natural salt exists. Herbivores also search for water plants in swamps, wetlands, or riverbanks because water plants contain more sodium than those grown on land. Ants that colonize near the ocean are fond of sugar, while ants that live far from the sea crave salt. On the ridges of Mount Elgon on the border of Uganda and Kenya, at an altitude of 2400 m, there is a cave that has been dug by generations of elephants, in order to find salt. Gorillas eat decayed trees in order to obtain sodium, which provides more than 95% of what they need. Lions eat the internal organs of prey first in order to obtain sodium. The carnivorous polar bears of Alaska enjoy eating seaweed during the seasons when seals are not available. All these animals exhibit these peculiar habits simply to consume salt.

The period between 500 and 1000 A.D. in Europe is known as the Dark Ages. The effects of global warming increased the sea level at the time. This elevation dramatically decreased the production of salt in saltern plots and, thus, created a scarcity of salt. This shortage created a series of deaths caused by dehydration and symptoms of mental instability. Such symptoms from salt deficiency were even worse inland. Eventually, people leaped around like mad men, and started to drink animal or even human blood in order to obtain salt because the blood of animals actually contains a certain amount of salt. Even today, in the upcountry of Africa, people stab the arteries of cows with sharp bamboo straws in order to drink their blood.

However, today, salt has become a very cheap material, and the level of daily salt consumption per person has grown to around 5–10 g. In fact, it is considered the main cause of many diseases and illnesses. If salt is so unhealthy for the body, why can't we easily cut out salt from our diet and reach a simple solution to our health problems? However, this is difficult because food would lose much of its taste without salt. Even though salt provides simple saltiness, it is the basic seasoning for many foods. Poorly seasoned food tastes bad, just as immature fruit that is not sweet does not taste good. It is essential that our diet consists of foods that are salty and have umami, and our snacks are sweet and sour. However, it is not just a problem of whether our food and snacks are salty or sweet or not. The problem is that without salt, all of the other taste elements and odorants would lose their highlights and taste would lose its balance because of the lack of salt. All tastes emphasize the saltiness as the main characteristic of their unique total taste. The decisive factor for determining the palatability of the food is the amount of salt. Expecting good taste from food that has not been seasoned properly is hopeless and thoughtless.

The role of salt in cooking is not merely to provide saltiness

Salt has many functions in foods other than providing saltiness. In addition to giving the salty taste to food, it has been used for preserving various foods through salting and fermentation control. Thus, the role of salt in controlling other taste and flavor changes will be described in this section. Here, we will reveal how experienced chefs control the overall taste of cooked food with salt.

The contrast effect

The contrast effect is when one taste element is felt to be stronger due to the stimulation of another taste. A good example of the contrast effect with salt is when the addition of a little bit of salt can strengthen the sweetness of sugar. A small amount of salt (around 0.5%) is added to sweet red-bean paste and watermelons to enhance the original sweetness. The addition of salt also makes meat broth and kelp soup taste better; this is also an example of the contrast effect.

The suppression effect

The suppression effect, also known as the retardation effect, is when the presence of one particular taste element weakens the strength of another taste. For example, the addition of a small amount of salt diminishes the sour taste of vinegar in vegetables. Japanese chefs use a slight amount of salt when preparing pickled plum and sushi in order to lower their strong taste of vinegar. The addition of salt makes them taste smoother. Another example of repression can be observed in fermented fish or fish sauce products. The reason these kinds of highly salted fish can be enjoyed, despite their high levels of sodium, is because amino acids and organic acids help to repress saltiness. A slight addition of sour taste to salty food can make the food taste less salty. On the other hand, the addition of salt to sour food can reduce the sourness. Salt can repress some undesirable tastes such as bitterness and other odors.

Furthermore, salt increases the volatility of odorants in proteins, which helps richen the flavor. In areas where salt is abundant, it is used more for preventing spoilage in foods than for seasoning. Especially in regions where fish is frequently consumed, salting to preserve seafood was the established tool for survival. Traditional food preserving methods, such as smoking or sun drying, also use salt. Before smoking or sun drying, fish are immersed in salt water. In the whole of history, fifteenth century Swedish people consumed the highest levels of salt. According to calculations, the amount of salt consumption per person in Sweden was 100 g per day because pickled fish was pretty much the sole source of food. Subsequently, the introduction of refrigerators substantially decreased the level of salt consumed. Typically, it is toxic to consume minerals at more than 3–5 times their recommended daily intake, but our body has a high resistance towards excessive salt consumption. Salt is used for many biological functions in the body, which include osmosis, dehydration, protein denaturation, and reducing the freezing point. Thus, salt is the primary food additive in the human diet.

Acidity monitors the biological metabolism

When carbon dioxide dissolves in water it becomes carbonic acid. The ocean is referred to as the home of all life because it provides the basic environmental conditions for life: carbonic acid and salt. Of course when carbonic acid exists with sodium chloride, it becomes sodium carbonate, which is a sodium salt of the acid that is found in sodium caseinate or monosodium glutamate (MSG). The human body has receptors that can sense carbonic acid, which can alleviate stress. The pH level of blood is constantly kept at a controlled level. No matter how much acid or alkali is ingested from our food, our blood maintains a constant pH level of 7.3–7.4. Even a pH level change of only 0.1 can cause a very dangerous life-threatening situation. Yet the majority of fruits are very acidic. Typical soft drinks have a pH level less than 3.5 and lemon juice and vinegar are also less than 2.5. However, regardless of how much of these acidic foods we consume, the pH of our blood does not change. What chemical substance in our blood stream allows our blood pH to be maintained at a stable level for long periods?

Out of all the acidic substances that are present in the blood stream, phosphoric acid (very important for energy metabolism and signal transduction) seems the most likely to be the strongest acid, and therefore, the most important acid for controlling the blood's pH level. However, based on the concentration of phosphoric acid, it does not seem to have a significant influence on the pH level. Instead, a weak acid, carbonic acid (H_2CO_3), and a conjugate base, bicarbonate ion (HCO_3^-), are abundant within the blood, and play central roles in buffering the pH change in the blood. Since the concentration of carbonic acid is highly dependent on the concentration of dissolved carbon dioxide in the blood, the major player in controlling our body's pH level is actually carbon dioxide.

Carbonic acid is one of the most basic organic materials and has existed since early on in the history of the earth. The origins of life have always been a topic for many scientific research projects. Miller's experiment on primitive atmospheres concluded that the major substances created in his flasks were mostly acidic amino acids, such as glutamic acid and aspartic acid, as well as some organic acids, such as formic acid, lactic acid, propionic acid, and acetic acid. The most abundant elements in our body are carbon, oxygen, and hydrogen. Carbon is the basic building block of all organic compounds. Hydrogen is the most numerous in terms of molecules. Finally, oxygen is the greatest in total mass, with a molecular weight 16 times that of hydrogen. The most abundant organic compounds in our body are lipids, carbohydrates, proteins, and organic acids. Lipids basically consist of carbon and hydrogen. Carbohydrates are composed of carbon, hydrogen, and oxygen in a 1:2:1, ratio, respectively. Most proteins are negatively charged in the particular pH range and function as weak acids. Organic acids contain more oxygen than carbon in their structure. Therefore, living cells and tissues, which mainly consist of carbon, oxygen, and hydrogen, have to be either at a neutral pH or slightly acidic.

Our stomach acid is technically hydrochloric acid, with sodium chloride is its main component. Stomach acid made from this salt sterilizes microorganisms and denatures proteins by lowering the pH level to 0.9–1.5. Acids are also created during the processes of fermentation and putrefaction of food. Fermentation and putrefaction are the same enzymatic or microbial reactions. The only difference between fermentation and putrefaction is whether or not the process is intentional. If the reaction is deliberate, it is called fermentation; otherwise it is putrefaction or spoilage. During both processes, sugars are decomposed into smaller molecules. Under anaerobic conditions, yeasts convert them into alcohols, however, with aerobic conditions, the alcohols are converted into organic acids, such as ethanol, which gives acetic acid. Many other types of organic acids are also created from this process, including lactic acid, citric acid, and gluconic acid. Therefore, microbial activities naturally increase the sourness of the food. In the past, when food was not as abundant, any food in a tolerable status had to be consumed rather than thrown away. The judgment procedure for identifying whether the food was good to eat (fermented) or not (putrefied) only depended on our sense of a sour taste and the odors, and such senses were a warning sign of danger. Sweet and umami foods are always safe to eat, but sour foods must be consumed with caution. Nowadays, various food safety procedures and regulations insure that we can enjoy extremely sour, acidic foods created by fermentation or through additives. For newborn babies, who lack any experience with food, sour foods are naturally rejected. Sourness and bitterness can be enjoyed after learning (through experience), but they are not naturally enjoyable as sweetness or umami are. Hence, sensing the presence of acid is used to alert us to the unhealthy condition of the food.

When glucose is metabolized, it becomes pyruvic acid. This process does not require oxygen. When the mitochondria have a normal oxygen level, the pyruvic acid is completely converted into carbon dioxide through the Krebs cycle. However, if there is a lack of oxygen, the pyruvic acid is converted into ethanol by yeasts, or into lactic acid by muscles or by lactic acid bacteria. Fermentation by-products and lactic acid inside the intestines act as effective inhibitory substances to the growth of microorganisms. Therefore, the claim that lactic acid bacteria are good for a healthy gut is approved by regulatory agencies. The lactic acid accumulated in the muscles is created mainly from anaerobic muscle movement for the immediate generation of ATP, not from fatigue. After the regular concentration of oxygen returns in the muscle, the accumulated lactic acid is utilized as a metabolite for ATP synthesis.

Meanwhile, sourness can increase the strength of various tastes and flavors, except for sweetness. Fruits must have sourness and sweetness to create their particular flavor. When fruits do not have sweetness, the flavor becomes weaker no matter how many odorants they contain. The same phenomenon applies to sourness. Without sourness, the flavor of fruits is not as apparent. Therefore, in order to maintain a balance of tastes, acid must be added. In addition, the different types of supplementary acids result in dissimilar sour tastes. The tongue detects the sourness of citric acid and tartaric acid quickly, but that of malic acid, lactic acid, and fumaric acid more slowly. Succinic acid could have umami. Among all the acidulants, acetic acid (vinegar) is considered the oldest additive. Its history can be dated to 3400 years before that of MSG. Nowadays, citric acids are also used as a common acidulants. Large amounts of citric acid are found in various citrus fruits. Compared with other organic acids, citric acid is used as the standard acidulant because of its smoothness and refreshing quality. As a key metabolic pathway for energy generation, the Krebs cycle is also known as the citric acid cycle or tricarboxylic acid cycle (TCA cycle).

Bitterness: if it's bitter, spit it out!

Why do children hate green bell peppers? Nowadays, green bell peppers have become relatively sweeter than the original wild types, thanks to breeding and domestication technologies. However, children still dislike them. In fact, it's not just children that dislike green bell peppers. Among all mammals, human beings are the only ones that eat peppers intentionally, while cows, horses, and even goats reject them. The bitterness of a green bell pepper is due to its alkaloids. The *Solanaceae* family, which includes potatoes, tomatoes, eggplants, chili peppers, and bell peppers, consists a lot of poisonous species that contain large amounts of alkaloids. Of course, we do not have to worry about being poisoned from eating green bell peppers because they contain such small amounts. Originally the bitterness was used by animals to detect toxins, in the same way as sourness was. Animals avoid eating bitter leaves by instinct because their bitterness is an indicator of poisonous chemicals. Children act in a similar way by rejecting all things bitter. The perception to bitterness is very sensitive until the age of ten and continues to desensitize with age. This is why it is very difficult for parents to persuade their children to eat bitter foods because it is hard for them to tolerate the bitter taste, and people with a greater number of taste buds tend to be more sensitive to bitterness. Nevertheless, human beings, compared with other primates, have undergone substantial desensitization towards bitterness.

There are more types of bitter taste receptors than there are for other tastes. There are only two umami receptors, one sweetness receptor, one sourness receptor, and one saltiness receptor, while there are 25 types of bitterness receptors. However, we recognize bitterness as just one taste, not 25 different bitter tastes. This simply means that this sense was not developed for enjoyment or to distinguish between a variety of different, delicate, bitter tastes. It is just a signal to spit out any bitter foods. Therefore, all animals seek sweet foods and avoid bitter foods. If someone experiences food poisoning after consuming a certain food, that person would naturally avoid that food in the future.

According to a recent study, even small worms can remember the taste of a harmful food and avoid eating it. A free-living (non-parasite) roundworm, *Caenorhabditis elegans*, is only 1 mm in length and moves underground to catch bacteria as food. However, along with the non-pathogenic bacteria in the soil, there are also some pathogenic bacteria, which can be fatal. A research team at the Howard Hughes Medical Institute at Rockefeller University in New York has conducted a research project on olfactory learning in *C. elegans* (Zhang *et al.*, 2005). They fed this worm pathogenic bacteria. After being exposed to the pathogenic bacteria, the worm modified its olfactory preference by avoiding the odors from the pathogenic bacteria and increased its preference for the odors from familiar non-pathogenic bacteria. The research project also answered their questions concerning neurochemical changes after the worm had been exposed to the odors from pathogenic bacteria, as a substance called Serotonin was released in its chemosensory neurons. This substance is related to the feeling of well being and happiness in animals and is secreted when their digestive organs become nauseous. The evidence of serotonin secretion confirms that the animal is in an uncomfortable situation.

There is no difference observed in the taste intensity of bitterness by children under nine. However, after going through their teenage years, females tend to be more sensitive towards bitterness and this becomes even more sensitive during pregnancy. This may be the instinct of motherhood to defend the fetus, because bitterness and astringency are potentially related to toxic substances.

Some people enjoy bitter tastes

Professor D. Norman at Northwestern University, Computer Science, stated that "The fact that human beings can enjoy bitterness represents that the culture and learning can govern their instinct" (Norman, 2004). People enjoy coffee, tea, and alcohol, which provide substantial levels of bitterness. However, once people learn that these foods are not poisonous, their rejection to them diminishes. For example, after several trials, the bitterness of dandelion or chicory salads can become an enjoyable compliment to fresh spring vegetables. As seen in this example, the instinct of rejecting bitter foods can be modified with training. Thus, Professor Norman is correct in saying that cultural environments reflect our ability to overcome our own instincts. If the bitter foods have any health promoting effects, they are enjoyed even more than just through training. An even more interesting example is seen in the entertaining culture of alcoholic drinks, where the drinks have an insignificant amount of essential nutrients. It is well known that one glass of alcohol before dining can help the appetite. However, a common alcoholic drink that people enjoy before dining is a Martini, which is a bitter alcoholic drink. In addition, in the English-speaking cultures, high sugar alcohol drinks are described with sweet in front of their name, while bitter alcohol drinks with low sugar content have the description of dry. In most cases, dry but flavorful alcoholic drinks are served with appetizers, while sweet alcohols are served with desserts. In the situations where selection is completely preference-based, we favor ordering the bitter drink.

There have been recent studies on why the bitter taste stimulates our appetite. Belgium Professor Inge Depoortere and his team proved that when food molecules with bitterness are bound with bitter taste receptors in the stomach, the hormone known as ghrelin, which stimulates the appetite, is released (Janssen *et al.*, 2011). It is normally thought that taste receptors exist only on the tongue, but, in fact, they are also scattered around the inside of the stomach and intestines. The receptors in our gastric tracks help the body to maintain homeostasis by continuously monitoring the relationship between the taste of food and the digestion conditions. Of course the sense from taste receptors perceived by the stomach and intestines do not reach the cognitive area of sensorial perception in the brain. Researchers concluded this result through animal experiments. The ghrelin level of rats that were injection-fed with bitter liquids into their stomach peaked (2.2 times) 40 minutes after the feeding. The amount of food consumed in 30 minutes by the rats with the injections was 20% greater than the amount consumed by the control group (rats injected with water).

If this result is transferred to our lives: drinking a Martini before dining stimulates ghrelin secretion and increases the appetite more than drinking cold water.

The reason we consume caffeine despite its bitterness

Nowadays drinking coffee is a common trend. But why do we like to drink coffee as adults, even though we hated the bitter taste of coffee as children? The rich odor of coffee may be an important factor that attracts us; however, decaffeinated coffee with the same aroma is much less popular. Therefore, caffeine can be identified as the most important factor for the preference of coffee. The level of adenosine increases in the brain when we are fatigued and this is one of the nucleic acids mentioned earlier in the chapter regarding umami. Adenosine is metabolized into caffeine through several reactions and, after being converted into xanthine and uric acid, ultimately breaks down into ammonia and carbon dioxide. When adenosine binds with adenosine receptors in the brain, the neurons become less active, and, consequently, this inactivity induces drowsiness and expansion of blood vessels in order to increase the blood supply in the body. Figure 2.6 shows the chemical structures of caffeine and adenosine and the structural similarity of them with dopamine.

Since caffeine is an intermediate substance metabolized from adenosine and is, therefore, similar to the molecular structure of adenosine, it can also bind with the adenosine receptors. However, caffeine cannot activate adenosine receptors. When caffeine binds with the adenosine receptor, it simply blocks adenosine from binding to this receptor. Thus caffeine and adenosine are competing binding agents. After the adenosine receptors are blocked by caffeine, even with an increased level of adenosine, drowsiness and blood vessel expansion are not induced since the receptors cannot initiate their signal processing.

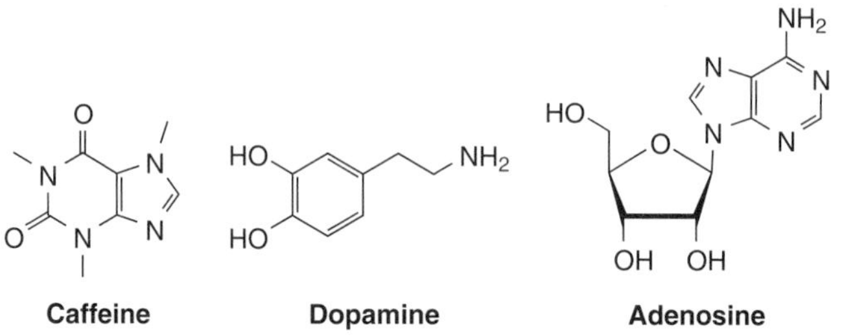

Caffeine   **Dopamine**   **Adenosine**

Figure 2.6   Chemical structures of caffeine, dopamine, and adenosine.

The blocking of adenosine receptors, therefore, sustains the contracted state of blood vessels, increases blood pressures, stimulates the liver to release blood sugar, and prepares for muscle contraction.

Caffeine also stimulates the central nervous system, slightly accelerating the heart and respiratory rates. This elevation excites the body before a certain task so that it can respond quicker. In addition, caffeine can also directly affect the muscles. Calcium has to be absorbed by the muscle in order for the muscle tissues to contract. Caffeine prevents adenosine from being absorbed into the muscle tissues, which causes more calcium to be rapidly released. Therefore, people who drink caffeine before working out tend to feel their muscles becoming stiffer. Furthermore, caffeine increases the secretion of dopamine, which stimulates the nerve cells. Because of these effects, caffeine is used as a cardiac stimulant, a respiratory stimulant, a central nervous system stimulant and a diuretic. Even though caffeine addiction is not as strong as drug addiction, it still has a similar addictive effect. Recently, scientists have found much evidence for the health benefits of coffee consumption. However, the side effects of habitual caffeine intake include more long-term consequences than short-term outcomes. For example, if someone tries to relieve fatigue by consuming caffeine, his/her body will demand more caffeine later. The constant increase in caffeine intake will also keep the body in a constantly stimulated state, which is not good for maintaining a healthy body. Adenosine receptors are important for sleeping, especially deep sleep, and there is nothing more important for health than deep sleep.

Caffeine does not make fatigue disappear; it simply covers it up. The main active ingredient in commercial energy drinks is caffeine. These drinks do not deliver energy or nutrients to relieve fatigue, but the caffeine simply blocks our body's sensing mechanism of fatigue. Caffeine can be better portrayed as a fatigue accumulator rather than a fatigue reliever. In the case of an emergency, it may be necessary to turn off the switch that identifies fatigue, but it is a good maintenance procedure to keep the switch on. The so-called energy bomb shots, made by mixing energy drinks with alcohol, which are popular amongst college students and office workers, are extremely dangerous. The sedative effect of alcohol and the stimulant function of caffeine nullify each other and make drinkers feel sober. However, in reality, the body is not digesting or getting rid of the alcohol more efficiently. Their minds are simply less sensitive to the intoxication. This makes the drinkers consume more alcohol beyond their typical tolerance limits. Masking the feeling of intoxication in order to drink more alcohol is very irresponsible. Enjoying lower amounts of alcohol is better for the body and the economy.

The olfactory sense is the dominant sensory perception of animals

Bacteria are single-cell microorganisms that are very small and only visible under microscopes. It may be hard to believe that single-celled microorganisms such as *Escherichia coli* have noses because of their size. *E. coli* does not have an actual nose, but, just as human beings can sense the chemical substances in the air with their nose, *E. coli* can recognize chemical substances with various types of sensors. The "nose" of *E. coli* is made of proteins and can recognize oxygen, amino acids, and sugars. The sensors are located at the tip of this rod-shaped microorganism. When the chemical substances bind to the sensors, chemical reactions take place on the other side of the protein sensors located in the cell membrane. *E. coli* recognizes asparagine acid, an important nutrient, and moves its flagellum toward the substance. *Paramecium* also has a similar sensing ability. These affinity reactions to chemical substances of microorganisms are called chemotaxis. The sense of microorganisms to any chemical is the equivalent sensorial mechanisms of olfactory systems in animals and, therefore, it is the most primitive sensory perception. The sense of smell was developed much sooner than any other sensory perception. It may have first appeared on earth 3500 million years ago.

For the majority of animals, the olfactory sensory has evolved into an essential instinct. Animals that are very sensitive to smell, such as dogs or squirrels, tend to walk around with their nose firmly placed on the ground, sniffing to pick up any odorants. The common feature of all mammals on the earth is that they have very well developed olfactory senses. However, as human beings started to walk upright, their olfactory senses diminished as they depended more on their visual senses. For birds that need to see long distances, sight is much more important and extremely well developed. Men depended on visual information for hunting, while women continued to have a relatively more sensitive sense of smell. In this section, we explain four important functions of the olfactory sense in the lives of animals: searching for food, avoiding danger, identifying others, and finding a mate.

The search for food

The most essential step towards survival is to find food, and the basic function of sensing odorants is to find this food. Insects are very good at picking up odors. Half of their brain cells are associated with the sensing of odorants. Mosquitoes locate animals and humans, in order to suck their blood, by detecting exhaled carbon dioxide. This shows how the olfactory sense can be used to find the origins of odorants. Rats also have the ability to detect the origins of odorants with just a few sniffs. Professor Bhalla's research team at the Indian National Center for Biological Sciences started a training program in which thirsty mice were granted water by determining the correct origin of the odors in a hole in the wall (Rajan et al., 2006). Surprisingly, the mice were able to determine the correct origin from the smell of bananas, eucalypti, or roses with an accuracy of 80%, regardless of what type of odorant was tested. The result showed that the mice only needed 0.05 seconds to determine the direction of the odorants. Moreover, hunger makes the olfactory sensory more sensitive. When hungry, ghrelin, an appetite-boosting hormone secreted by the stomach, stimulates the brain region in charge of delivering sensory information. For animals and human beings it has been revealed that this stimulation improves the olfactory ability to detect an energy source. During the time of the dinosaurs, most mammals were nocturnal. In order to find food and to avoid danger without any light, the sense of odorants was much more important than a visual sense.

Avoid danger!

Our memories and perceptions of negative stimuli are much stronger than those of positive stimuli. If someone followed a pleasant smell and found delicious food, that person would remember the odor and its reward very well. However, if someone followed an unpleasant odor and arrived at a lion's den, that unpleasant odor would never be forgotten. It is very important to survival that we remember dangerous situations more than pleasant situations. Animals can avoid their enemies with their sense of smell and they can even convey fear through their olfactory sense.

It has been discovered that an odor is given off by mice and fish during moments of fear and that odor induces fear in other mice and fish. In 2008, the researchers from University of Lausanne in Switzerland discovered the fact that mice give off an alarm pheromone, when they encounter danger (Brechbühl et al., 2008). The pheromone signaled other mice to run away or hide from the odor. When the air in the chamber of euthanasia of an old mouse was collected and injected into a different chamber of test mice, they immediately ran away to the other side or froze in fear. However, the mice whose Grueneberg ganglion was removed did not react any differently. The Grueneberg ganglion, a nervous tissue, was first discovered in 1973, but its exact function remained unknown until this study. In this experiment, the removal of the Grueneberg ganglion of mice did not affect the mice's ability to find hidden food. It was concluded that the removal of the Grueneberg ganglion only affected the mice's ability to sense the danger hormone and did not affect their ability to smell. Similarly, it has been revealed that fish also release screaming substances from their skin when attacked by bigger fish, which inform other fish of potential dangers.

Know who it is!

For animals, the sense of smell is the dominant sensory perception and also an essential method of communication. There is a difference in odors both between and within species as well as individual variations. Human newborn babies recognize their mothers' scent through their unique odors. Animals can also recognize those of the same race or the same family. It is also well known that animals spread secretions to mark their territory. This is an attempt to prevent other males from approaching and to display their territorial power.

It is said that odors can determine a tribe's identity in the rain forest. For example, the Desana Indians in the Amazonian tropical forest of Colombia have a unique odor different from all tribe members because of their genetic characteristics and the food they eat. These hunters often give off the musky smell of their food. Similarly, the neighboring fisherman tribe of Tapuya gives off the odor of fish. The neighboring Tucano tribe gives off the odor of root and bulbous plant vegetables grown in their fields.

In between cultural boundaries, the unique odor of traditional foods becomes an invisible barrier. In nearly every culture, there is at least one food with a disgusting odor. Surströmming in northern Sweden is a great example. Surströmming is fermented Baltic herring, which has a disgusting odor even to the Swedish people, who consider it to be a great cuisine.

In Scandinavia, there is a dried fish called lutefisk in Norwegian (lutfisk in Swedish or ludfisk in Danish). Lutefisk is naturally dried cod or whitefish that has been soaked in water for a week, immersed in caustic water for two days for protein gelatinization and lipid saponification, then put in plain water for a few more days. Garrison Keillor wrote in his book *Lake Wobegon Days* that "Every Advent we entered the purgatory of lutefisk, a repulsive gelatinous fishlike dish that tasted of soap and gave off an odor that would gag a goat." However, people who regard themselves as genuine descendants of Cnut or Vikings eat lutefisk at least once a year. It represents an affiliation to their culture. Similarly, a person who does not eat fermented chou tofu, is not considered to be a true Taiwanese. In Japan, they eat natto, which is a fermented bean paste, and in Korea they have fermented skate fish (i.e., thornbay ray). Nowadays, most nationwide franchise restaurants remove all characteristic tastes; however, the best way to welcome people who have returned home from foreign towns or areas is by sharing the local fermented food with their unique tastes and flavors.

Find a mate!

What kind of sensory perceptions are most important in finding love? The majority of people would think that the visual sense is the most important. In most cases, the primary factor for feeling a favorable impression towards the opposite gender is the appearance. The human brain has developed to make us depend more on our visual sense. However, for animals, the olfactory sense can be regarded as the most important factor in finding their mates. Odors of the opposite gender also play an important role in attracting the mate. The odor mark is used to exhibit their territorial areas. Human beings are also influenced by odors, just like animals, regardless of whether they understand the influence of the odor or not. This is because the limbic portion of the brain that controls our emotional sensitivity plays an important role when we are recognizing an odor. When butterflies look for their mates, the male can fly more than 10 km towards the exact location of its female partner based on the trace amount of pheromone that is released by the female. Even most mammals, like dogs and deer, are excited by the odor of their partner's reproductive organs before mating. The word pheromone is actually constructed from the Greek words *pheran* (to transfer) and *horman* (to excite). Pheromone refers to the chemical substance used for communication between species. For animals, according to their functions, pheromones are classified into aggregation, alarm, releaser, signal, development, territorial, trail, information, and sex.

The dominating influence of the olfactory sense can be easily understood by considering salmon swimming upstream over waterfalls. In order to find a safe place for breeding, salmon seek the fresh waters upstream, away from the more competitive environments in the oceans. They swim upstream by following the traces of the same odor of water that they remember when they were hatched. Even though their bodies are capable for living more than 10 years, swimming upstream uses all of their energy, reducing their lifespan to 7 years. This energy burst is triggered simply by the olfactory sense.

Thus, this chapter is only an introduction to the most basic functions of the olfactory sense. Setting aside the more detailed roles for a later chapter, the relationship between taste and flavor will be explained in the next chapter.

References

Brechbühl, J., Klaey, M., Broillet, M.C. (2008) Grueneberg ganglion cells mediate alarm pheromone detection in mice. *Science.* **321**(5892): 1092–1095.

Covington, D.K., Briscoe, C.A., Brown, A.J., Jayawickreme, C.K. (2006) The G-protein-coupled receptor 40 family (GPR40-GPR43) and its role in nutrient sensing. *Biochem. Soc. Trans.* **34**(5): 770–773.

Hayes, J.E. (2008) Transdisciplinary perceptives on sweetness. *Chemosens. Percept.* **1**(1): 48–57.

Janssen, S., Laermans, J., Verhulst, P.J., Thijs, T., Tack, J., Depoortere, I. (2011) Bitter taste receptors and α-gustducin regulate the secretion of ghrelin with functional effects on food intake and gastric emptying. *Proc. Natl. Acad. Sci. USA.* **108**(5): 2094–2099.

Norman, D. (2004) *Emotional Design: Why we Love (or Hate) Everyday Things.* Basic Books, New York.

Pepino, M.Y., Love-Gregory, L., Klein, S., Abumrad, N.A. (2012) The fatty acid translocase gene CD36 and lingual lipase influence oral sensitivity to fat in obese subjects. *J. Lipid Res.* **53**(3): 561–566.

Rajan, R., Clement, J.P., Bhalla, U.S. (2006) Rats smell in stereo. *Science.* **311**(5761): 666–670.

Zhang Y., Lu, H., Bargmann, C.I. (2005) Pathogenic bacteria induce aversive olfactory learning in Caenorhabditis elegans. *Nature.* **438**(7065): 179–184.

3 Taste is General Science

Pour moi, je suis non seulement persuadé que, sans la participation de l'odorat, il n'y a pas de dégustation complète, mais encore je suis tenté de croire que l'odorat et le goût ne forment qu'un seul sens, dont la bouche est le laboratoire et le nez la cheminée, ou, pour parler plus exactement, dont l'un sert à la dégustation des corps tactiles, et l'autre à la dégustation des gaz … On ne mange rien sans le sentir avec plus ou moins de réflexion; et pour les aliments inconnus, le nez fait toujours fonction de sentinelle avancée, qui crie: Qui va là? Quand on intercepte l'odorat, on paralyse le gout … …

For my own part, I am not only persuaded that without the interposition of the organs of smell, there would be no complete degustation, and that the taste and the sense of smell form but one sense, of which the mouth is the laboratory and the nose the chimney; or to speak more exactly, that one tastes tactile substances, and the other exhalations … We eat nothing without seeing this, more or less plainly. The nose plays the part of sentinel, and always cries "WHO GOES THERE?" Close the nose, and the taste is paralyzed …

Jean Anthelme Brillat-Savarin (Physiologie du goût, 1825)

As stated in the previous chapters, there are five tastes: sweetness, umami, sourness, saltiness, and bitterness. The real flavors perceived in apples, strawberries, watermelons and all other foods are their odors rather than their tastes. The only tastes that we sense from fruits are sweetness and sourness. Without a doubt, there are many different factors that influence the unique taste of a food, such as color, shape, and texture. However, the diversity of flavors only means that there is a large variety of odors, not a large variety of tastes.

How Flavor Works: The Science of Taste and Aroma, First Edition.
Nak-Eon Choi and Jung H. Han.
© 2015 John Wiley & Sons, Ltd. Published 2015 by John Wiley & Sons, Ltd.

Some people may think that the taste and flavor of an apple can be sensed more strongly when chewed in the mouth. However, from the five tastes, apples only have sweetness and sourness; and with just these two tastes, it is impossible for us to differentiate an apple from another food. Taste does not significantly influence the holistic flavor. The distinctive flavor of the apple is due to its unique odor profile. In fact, during the oral process of chewing an apple, its odorants volatilize through the narrow air passage behind the throat that connects to the nose. The odorants are then associated with the two tastes of the apple and produce the final flavor of the apple. Thus we perceive the flavor of the apple by smelling it through our mouths. If we closed our eyes and noses, and only relied on taste, it would be very difficult to distinguish an apple from an onion. Tastes play only 10% of the role in determining flavor, while flavors are the other 90%.

Is there a difference between the flavor of apple cider produced by completely grinding the apples and the flavor of fresh apples? How about the difference in tastes between apples stored in a refrigerator and those stored at room temperature? Taste can change depending on temperature, texture, color, and shape of the food as well as with expectations, the sound made by chewing, physical/emotional conditions of the consumer, culture, presumption, and experience. Taking these many influencing factors, which are shown in Figure 3.1, into consideration, the mutual relationship between taste and flavor will be discussed in this chapter.

Taste improves with harmonized combinations

Tastes can be largely classified into two categories: meals (i.e., main dishes and staple foods) and desserts. In Western cultures, snack foods are a salty food category consumed between meals; while in Eastern cultures, snacks are generally sweet and are categorized as desserts, even though they are not consumed with regular meals.

The taste of meals = saltiness + umami + savory flavor

When a food is not tasty, the reason is generally due to the lack of salt. Salt is the most important and basic seasoning for creating the taste of foods. An understanding of the basics of tastes is achieved by considering why salt is so important in creating taste. It has the ability to lower any undesirable flavor and sourness. The other taste that harmonizes

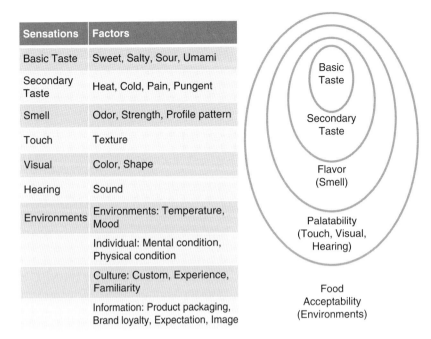

Sensations	Factors
Basic Taste	Sweet, Salty, Sour, Umami
Secondary Taste	Heat, Cold, Pain, Pungent
Smell	Odor, Strength, Profile pattern
Touch	Texture
Visual	Color, Shape
Hearing	Sound
Environments	Environments: Temperature, Mood
	Individual: Mental condition, Physical condition
	Culture: Custom, Experience, Familiarity
	Information: Product packaging, Brand loyalty, Expectation, Image

Figure 3.1   Factors for food preferences.

well with saltiness is umami. Foods such as fermented soybean stew, fish stew seasoned with soy sauce, udon (Japanese hot noodles), and mushroom soup would not have the same pleasant taste if they did not have umami. These two tastes (saltiness and umami) are the foundations of all foods in meals and savory snacks. Sweet and bitter tastes are added from time to time in order to create a deep and rich taste, but it is not very common in everyday meals. We take much pleasure from saltiness and umami because they have a savory flavor.

The taste of dessert (and fruit) = sweetness + sourness + sweet odor

Fruits that are not ripe do not taste good because they lack sweetness. Sweetness is essential for a good taste in snacks and desserts. If the first requirement is sweetness, the second requirement is sourness and the third is harmonized odors. These must be fulfilled for desserts and sweet snacks to taste appropriately. This is why the most critical condition in the formulation of soft drinks is the ratio of sugar to acid (i.e., ratio of Brix to acidity) along with the right flavors.

When 2–3 tastes are combined together in a food, the strength of each taste is more enhanced at the lower concentration. For example, saltiness increases the strength of sweetness. The addition of a small amount of salt can enhance the sweetness of table sugar. Saltiness also enhances the strength of umami and other flavors. Similarly, umami also enhances saltiness. By adding MSG, the amount of salt can be lowered by 30% whilst keeping the same level of saltiness. MSG is sometimes misconstrued to be an artificial seasoning. It is also sometimes mistaken to be an artificial substance, and, as such, is treated as if it were a monstrous chemical. Therefore, the increased umami caused by only a pinch of MSG raises much apprehension among consumers. The enhanced taste in foods does not come solely from the increased umami due to the MSG, as a variety of other tastes are required for the addition of MSG to have any positive effect at all. The addition of umami does not make the food taste better, but the enhancement of the strength of all the other tastes caused by the addition of the MSG does. MSG provides the umami and enhances other tastes. All taste elements are interrelated and should be harmonized to achieve the best overall taste.

Tastes influence odors

Aroma, the sense of odorants, is ultimately responsible for the variety of different tastes perceived when eating. Soap, with oleic acid as its main ingredient, would be hard to be distinguished by taste alone if the nose was blocked. Ferrous sulfate is hard to recognize just by its taste and without the nose's sense of its metallic odor. However, non-volatile substances such as MSG cannot be recognized by their odor and to identify them, they must be tasted.

It is sufficient to create a good taste by combining the sweetness in a moderate amount of table sugar or sweeteners, sourness from 0.1% of acidic seasonings such as vinegar or citric acid, saltiness from 0.9% of salt, and umami from 0.4% of MSG. The preference for a food can be determined by 10% from the taste and 90% from the variety of odors. However, the standard criterion for choosing food depends on its so-called taste, not its odors. It is hard to find a situation where bad tasting food sells well. However, since the flavor of the food is dependent on its odorants, it can be said that the overall taste of a food is also dependent on its odor, even though the odor has no meaning by itself without the foundation of a good sense of taste. A good overall taste cannot be achieved by just a good odor alone. Taste has to be the foundation of the overall taste before the addition of any odorants.

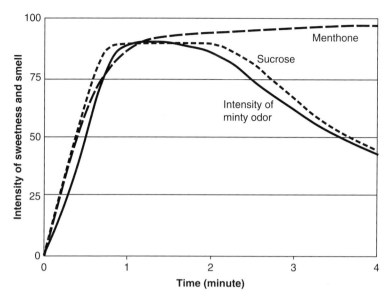

Figure 3.2   The effects of concentrations of sucrose and menthone on the intensity of minty odor. Modified from Taylor *et al.* (2000).

The intensity of an aroma (i.e., the olfactory sense) can be influenced significantly by the taste elements. For example, sweetness and sourness can help make the smell feel stronger. In the graph shown in Figure 3.2, the relationship between the sweetness and the intensity of the smell is displayed in a simple manner for a chewing gum containing a minty aromatic chemical, menthone, and sucrose. Within 2 minutes of starting to chew the gum, the concentration of sucrose decreased. During this period, the actual concentration of the menthone was still maintained at a high level, but the consumer felt that the intensity of its minty odor had dramatically decreased. The intensity of the smell seems to be influenced significantly by the sweetness more than by its concentration. Therefore, to make the smell last longer, the key strategy is to find a way to maintain the period of the sweetness sense rather than the duration of the olfactory sense. This is a clear and simple example, which shows that our sensory perceptions are inter-related and inaccurate.

A similar result was reported when combining a sweetener with odor and acid. This test was carried out with strawberry odor, sweeteners, and acidulants. The strawberry odor was supplied, blocked, and resupplied in 2 minute time intervals. In the first experiment, incorporation of the sugar and the acid was maintained, while the supply of strawberry odor was blocked. Surprisingly, the subject did not notice the sudden absence of strawberry odor. However, when the acid was blocked and the levels of sucrose and strawberry odor were maintained, the subject thought that the strawberry smell became weaker by 40%. The additional removal of sugar from the samples, but still maintaining the original levels of the smell, decreased the intensity of the scent even more severely. The subject felt that the strawberry scent had completely disappeared. The removal of both acid and sugar made the subject unable to sense the smell at all.

As you can see, the function of the odorant is only effective with the presence of a taste foundation. Of course, the sudden removal of the odor or a very high concentration of the odorant would be noticeable after a long period of time, but within the general conditions of the average concentration of the odorant in foods, the taste is an essential foundation to build the overall taste preference. The presence of odor without any good taste is like the presence of a king without any subjects.

Food has to be dissolved for us to taste and chewed to enhance the taste

Odor can only be sensed when odorants from the food volatilize and reach the olfactory receptors. Therefore, the perception of the odorant is not just the odorant itself, but also the physico-chemical conditions of the food that contains the odorant (due to the interaction between the food components and the odorant) and on the release of the odorant through the chewing mechanism. Therefore, it is appropriate to evaluate the smell with the food and not just with the odorant itself. When odorants are added to food, the smell of the food can change depending on the influences of its cooking/manufacturing processes and the interaction of the odorant with various food components. There can also be a difference in the release profile of the odorant depending on the degree of chewing, chewing time, saliva, and so on. We tend to chew food several times over a period of around 20 seconds in order to sense the taste before swallowing it.

The tongue has a total of eight muscles. The lack of bones in its muscle structure makes the tongue naturally flexible. This flexibility of the tongue means it is possible to close the entrance to the windpipe when we need to hold water inside our mouths. When the tongue holds food within the mouth cavity, the chewing process is activated to reduce the size of the food. Once the food has been adequately chewed, the tongue starts to blend the food inside the mouth. The teeth and jaw imitate the actions of a dough mixer. Finally, once the food is chewed completely, the tongue is also responsible for determining whether it is time to swallow it or not. The somatosensory perceptions of the tongue determine whether the food still contains any solid particles that might be hard for the stomach to digest. When the tongue finally determines that the chewing process is complete, it opens up the throat and starts the motion of swallowing. It is also the role of the tongue to transfer the food into the throat. In a wave-like motion, the tongue quickly pushes the whole contents of the mouth into the throat.

The delicate movements of the tongue are only a portion of its remarkable functions. It also plays a role in tasting the food for enjoyment during the chewing process and helps the taste buds on the surface of the tongue to sense the tastes of the food. Other than the five basic tastes, it can also perceive around 20 other types of senses, such as temperature and texture. In fact, the temperature of the food is always an important part of its taste. The texture, including the smoothness, hardness, stickiness, crispiness, or chewiness of the food, also makes a significant contribution to the overall taste. The effect of texture on the overall taste can be easily observed for the tastes of a food before and after grinding.

The main ingredients influence taste and odor

Even though carbohydrates, proteins, and lipids do not actually have any taste, they significantly influence the overall taste of the food. This is especially true in the case of lipids. The release of odorants depends on their solvents, typically water or oil, and most food products will contain certain amounts of lipids. Most substances providing a scent are lipophilic and dissolve in oil, but with some degree of hydrophilicity. Therefore, the smell of the food is highly dependent on the lipid content. The lipids in the food capture the odorants and slowly release them. Depending on the lipophilic nature of the odorant, the level of lipid solubility and release kinetics of the odorant can be changed. This can cause differences in the odor spectrum, but from an overall consideration, the release rate of the smell becomes more gradual and slower on combination with the lipids in the food. Therefore, the odor of the food can become smoother and richer. Thus, when all the lipid ingredients are extracted or excluded from the food, the overall smell of the food is completely changed. New product developers must consider the effect of lipid content on the flavor profile of non-fat or reduced-fat foods. The higher the lipid content of the food, the lower the amount of the odorant released, and, therefore the concentration of the odorant has to be increased.

Carbohydrates have less of an influence on odors than proteins and lipids. Monosaccharides and disaccharides have almost no effect on the release of odorants, but influence the odor recognition. Generally, the increase in sweetness intensity helps with the recognition of the odors. For example, unripe fruits do not enhance the odors, resulting in bad taste. Since the molecular weights of extremely sweet artificial or natural sweeteners are very low, they do not influence the release kinetics of the odorants. However, they can alter the odor. A sweetener such as aspartame has very low molecular stability, meaning that the level of sweetness is easily changed. Aspartame can be readily degraded because it consists of only two amino acids, especially with aldehydes. Once the structure of aspartame has changed, its sweetness disappears, and, consequently, the smell of the food also changes.

Starch has a high molecular weight and influences the physical chemistry properties, such as the viscosity, of the product. A high viscosity makes the odorant less volatile and delays the time the odorant requires to reach the olfactory receptors. Since starch has a less hydrophilic part in its helix structure, it can entrap the odorous chemicals. Besides starch, cyclodextrin also has a molecular structure that can entrap the odorous chemicals. Starch also has the ability to entrap the chemicals, but the entrapping ability of cyclodextrin is greater than that of the starch. There are some modified starches that possess a better ability to entrap the odorants, which is achieved through the modified functional groups. These are used as adhesive coating materials for powdered flavors. The thickening polysaccharide and gelling carbohydrates are used in jelly production to make the structure of the product firmer and, consequently, lower the release rate of the odorants. Therefore, this issue must be resolved by increasing the concentration of the flavoring agents. Not only does this interfere with the perception of the flavors, but it also interferes with the sense of sweetness, by lowering it. Amongst other carbohydrates, gellan gum and CMC (carboxymethyl cellulose) have the lowest interference effects on the release of the flavoring chemical.

Unlike carbohydrates, proteins have various sites to bind to the odorants through hydrophilic, hydrophobic, and electrostatic interactions. Therefore, based on the composition of the amino acids, proteins can change the binding forces from weak to strong through the odorants. Since proteins have very complicated three-dimensional structures that are unstable with respect to the environmental conditions of pH, salt, and heat, their interaction with various odorants are not very predictable. When the odorants bind with the —OH, —NH$_2$, and —SH groups of proteins, their interactions become very strong and the release of the odorants becomes much slower. The overall olfactory profile and spectrum will be changed.

When we categorize the binding force of proteins with common odorants, the order of the binding force from strongest to weakest is soy protein, gelatin, egg protein, and milk protein. Therefore, when a flavoring agent is added to a product that contains soy protein, due to the combination of the odorant and the protein, the release kinetics of the odorant are retarded. Because of the difference in the binding capacity of the odorant to proteins in the food, it is hard to recreate foods with identical flavors. The structure of the protein changes with respect to pH and then these structural changes also affect the flavor profile. When the protein is denatured by high heat, the increased hydrophobicity means the protein easily binds with the odorant. The protein also affects the physical properties of the product and such changes then influence the perceptions of taste and odor.

Sound has an influence on taste

The perception of hearing also has an influence, albeit a very low one, on the taste of foods. The crunchy sound heard when eating potato chips is one of the important elements indicating the total quality of the chips. In an experiment, a subject was given a certain type of potato chip. While the subject ate those chips, a recorded sound made from eating a different type of chip was played; the subject was unable to distinguish the difference between the two potato chips. Similarly, vegetables and fruits with a crunchier sound receive better preference scores. The crunchy sound of fruits indicates how ripe they are. The frying sound of deep-fried foods and the pouring sound of wine can increase the expectations and appetite for the foods. Therefore, all of our senses directly influence the perception of the taste. It is impossible to understand the overall taste of a food through just one of our separate senses.

Visuals, colors, and food styles

It is true that visually appealing desserts are more delicious. Visually prettier foods stimulate our appetite further and produce more saliva in our mouth. Saliva, acting as a digestive enzyme, also helps the tastes buds to make contact with the food in an even manner during the chewing process.

There are lots of photographs of fruits dyed with blue coloring on the Internet, which simply eliminate any appetite for these foods. Color can completely change our perception of taste. The reason that blue colored fruits do not look good is because the majority of the natural food colors are yellow, orange, or red. Our appetite is very conservative, so it is extremely hard to find a blue food appetizing because of its unnatural color. The color has to harmonize with the taste and the odor of the food for an overall pleasing taste. Our visual sense provides a lot of information to help distinguish the taste of a food. Differentiating a dark cola from a colorless soft drink is not a hard task thanks to the visual cues. In a blind test with our eyes closed, the majority of us would be able to tell the difference because of their completely different flavors. However, our senses can be easily tricked. Whilst blindfolded, subjects were given two drinks of the same cola samples, one after the other, and then asked which one was the real cola. These subjects did not answer that both were the same. Most answered that one was cola and the other was the colorless soft drink from before. When red food colorant is added to white wine to make it look like red wine, even wine specialists have a hard time determining the difference. If strawberry flavored candy was colored yellow and banana flavored candy was colored red, it would also be very difficult to identify that the colorant had been switched. Since they both have familiar fruity scents, the red candy is thought to be strawberry flavored even if the real flavor is banana.

It was thought that mammals used to have the ability to perceive various colors utilizing types of G-receptors. However, during the dinosaur era, the majority of mammals had a nocturnal lifestyle, and, therefore, their perceptions of odor and sound developed more, while their color perception deteriorated to only two types of photoreceptor cells for perceiving vision. The ability to perceive various colors in detail, like humans, is not common for other animals, although early primates had photoreceptor cells that were only sensitive to two colors: blue and green. Therefore, the ability to distinguish the color red has developed relatively recently.

Why, then, did primates evolve a method of distinguishing the color red? Scholars claim that the ability to see red in a green jungle would give them many benefits because the primates could look for fruits as an important food source. The capacity to see red does not mean the simple addition of the ability to distinguish the color red. It means that they can see the overall visible wavelength spectrum. This visual ability provided animals with a method of identifying whether the vegetables were still in their unripe stage or fully ripened. There are self-defense phytochemicals and indigestible hemicelluloses in plants that may cause problems to the digestion systems of animals. The more mature leaves have more anti-nutritional substances. Thus it was necessary for herbivores to have the ability to distinguish between the fresh new leaves as their food and the mature ones.

Color creates a sort of expectation of something special from the food. When a colorant is added to only one sample solution out of two odorless solutions, people tend to think that the colored solution has a flavor, and the colorless solution has no odor. When both of the solutions contain the same concentration of an odorant, people tend to think that the one with color has a stronger smell. However, this result only applies when the test is by inhalation through the nose. On the other hand, when the solution enters the mouth, it is perceived that the solution sample without any color has a stronger smell than the colorless sample. This shows the extensive degree of interaction between the complexity of our brain and the sensory organs. Unpleasant smells are perceived with a greater intensity when they are sensed by the mouth rather than by the nose. The main purpose of our olfactory system is not to find foods, but to determine whether the food found by the visual system is good to eat. It must have been helpful for survival to be able to objectively determine the unpleasant tastes of unknown foods.

Why does color exist?

The sun produces around 3.9×10^{28} J s^{-1} of energy, which is equivalent to the total energy of one quadrillion (10^{15}) nuclear bombs. The nuclear fusion energy is released in the form of gamma rays or neutrinos. The plasma light produced by the nuclear fusion in the core of the sun travels to the surface and is released into space. It takes about 500 s for this light to reach the earth. Thus the sunlight we see now originated from inside the sun a long time ago, ranging from several thousand to ten million years.

The inside of the sun is a high-pressurized plasma state, ionized from hydrogen gas. The light from the sun is produced at the core by nuclear fusion of this hydrogen gas. This process is repeated within the sun innumerable times before the light finally escapes to the surface. It also gradually changes to a longer wavelength. By the time it finally reaches the earth after passing through the ozone layer, the gamma rays of the sun have become the wavelengths of visible rays. This is the reason why all life forms can continue to flourish on the earth as organisms in the plant kingdom have the ability to photosynthesize. Photosynthesis is the series of reactions that produce organic materials using the visible rays of sunlight. Porphyrin group chemicals, such as chlorophyll, and the carotenoid group, such as lycopene or carotene, are the major natural colorants. Once these natural colorants had been included in color receptors, animals obtained their visual sense. It took, at the most, a million years for the color receptors to evolve completely into the eyes. The visual sense is mainly utilized to find foods. Along the broad light spectrum, humans can only perceive a small range of wavelengths, around 0.4–0.7 μm, as color, which means we can perceive an extremely narrow range of light. However, it is actually more efficient to be able to only perceive this narrow range of light with high definition rather than a wider range with poor definition. Most information is collected by the visual senses. This is essentially an incredible amount of information.

The green chlorophyll, red hemoglobin, and yellow vitamin B12 are actually part of the same chemical group, porphyrin (Figure 3.3). The chemical structures of porphyrins are made from glutamic acid. The addition of a magnesium ion (Mg^{2+}) to the porphyrin produces chlorophyll, the addition of ferrous ion (Fe^{2+}) gives hemoglobin, and the addition of cobalt ion (Co^{2+}) gives a yellow colored vitamin. Hemoglobin, with a ferrous ion (Fe), has a red color, as seen with oxygen in the arteries, and a blue color, with carbon dioxide in the veins. The process of heating hemoglobin oxidizes the bonds with oxygen and causes browning, but most people prefer meat not to be brown. However, as hemoglobin interacts strongly with cyanide (CN), carbon monoxide, and nitric oxide (NO), the addition of these chemicals prevents oxidation of hemoglobin and thus maintains the pinkish red color of the meat. The addition of nitrite also stabilizes oxidation and helps maintain the red color in cured meats such as hams and sausages. In addition, the binding of hemoglobin with hydrogen sulfide (H_2S) creates a green color. It would be very strange if the color of blood was green, but color changes are created by the apparently slight differences in the chemistry.

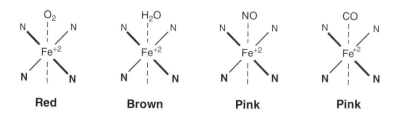

Figure 3.3 Colors can completely change depending on which molecules bind to the center of hemoglobin.

The basic structure of pigment: why are there no naturally blue foods?

Organic pigments can absorb the long wavelength of light if the number of conjugate bonds increases. The pigment must have at least 11 conjugated bonds consisting of repeated saturated and unsaturated bonds. Thus, it must have a carbon number of at least 22 to display color. While the number of the carbons in fatty acids is normally less than 18, carotenoids are made up of 40 carbons. Pigments are generally of two types: carotenoids and flavonoids (Figure 3.4). The carotenoid group has a base structure of lycopene. Since it does not have unstable side groups, which are essential for exhibiting its color (i.e., an auxochrome side group), it maintains a consistent red color regardless of pH change and temperature. However, the pigments in the group are lipid soluble and weak with respect to oxidation. Therefore, oxidation has to be prevented by blocking light or oxygen and by providing sufficient antioxidant vitamins, such as vitamin C or E. Carotenoid has a number of diverse roles in plants. One of these roles is to assist photosynthesis within the chlorophyll. It absorbs the wavelengths of light energy that the chlorophyll cannot absorb, and

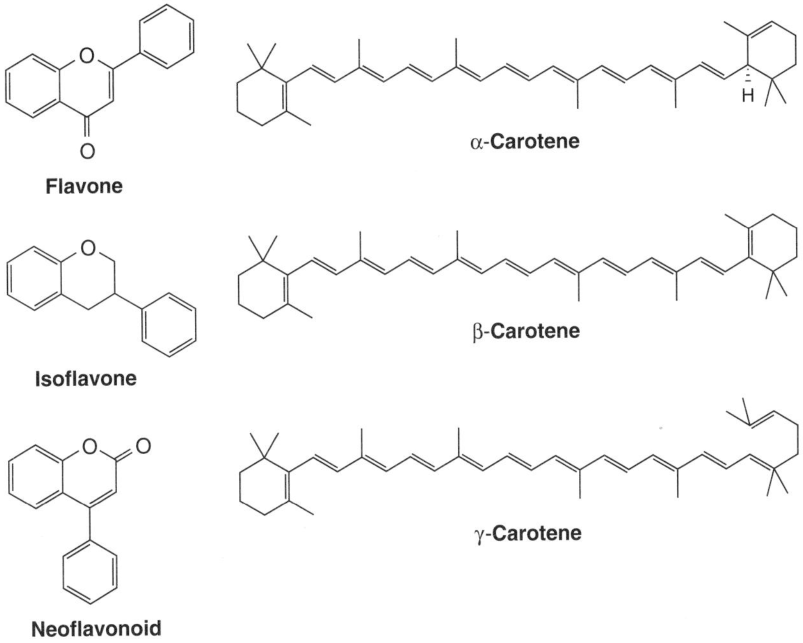

Figure 3.4   Various flavonoids and carotenoids.

delivers this light energy to the chlorophyll. It is oxidized in place of the chlorophyll in order to protect it.

The pigments of the flavonoid group have the basic structure of flavone. Flavonoid itself is lipid soluble, but once it combines with sugar it becomes a water-soluble anthocyanin. This has led to more than 600 kinds of anthocyanins being studied for their functions as antioxidants. Flavonoid is created from phenylalanine, an amino acid. Initially, the many health benefits and biological functions of flavonoids were thought to be due to their antioxidant effects. However, after numerous tests, it was revealed that the biological functions of flavonoids are caused by their work in the cellular signaling system. The concentration of flavonoid required to activate the cellular signaling system is much lower than that needed for the anti-oxidant effect. Also, flavonoids are found in a symbiotic relationship with nitrogen-fixing bacteria in the root tubercles of leguminous plants. Anthocyanins in red maple tree leaves inhibit the growth of other plants after decomposing in the soil. They also show antimicrobial effects, especially in the legume family. In conclusion, flavonoids function as a type of hormone for plants and because of this they are not always beneficial for our health. Moreover, the auxochrome side groups are very sensitive to pH, and consequently can dramatically change the color of flavonoids, thus they are not very convenient for use as food colorants.

Unfortunately, an ideal pigment does not exist in the natural world. The molecular size of a food pigment is fairly large, so its solubility is low. Carotenoids are mostly lipid soluble and weak against oxidation, and flavonoids are slightly more stable against oxidation but much more sensitive to pH. Synthetic pigments can overcome such problems but are avoided due to the nonscientific controversies surrounding their safety. Because of the fact that only a very small amount of a synthetic pigment is need to create a strong color, it is easy to be under the assumption that synthetic pigments are very strong substances and harmful. However, synthetic pigments possess ring structures, and several of these are combined in order to stabilize against oxidation. Even though they have high molecular weights, they contain sodium in order to increase their solubility.

Compared with synthetic pigments, natural carotenoids such as lycopene have simpler chemical structures and lower molecular weights. Thus, the color of the lycopene is three times stronger than the same concentration of a synthetic red colorant. Lycopene is the major contributor to the red color in tomatoes, even with only 0.002–0.008 g for every 100 g of tomato. Tomatoes can be perfectly red with only a small amount of its natural red pigment. In the past, 12 000 clams had to be caught from the Mediterranean Sea in order to obtain 1.2 g of Tyrian purple and in order to obtain 1 kg of crimson red cochineal carmine dye, 10 000 female cochineal insects had to be captured. Thus there are natural methods of creating strong colors with low concentrations. The reason why we use large amounts of natural colorants to obtain strong colors is due to the low purity of commercial natural pigments in the colorants. Thus it is very difficult to use natural pigments to express desired colors.

Approved food grade synthetic pigments are safer than natural colorants. There are countless numbers of stories on the Internet claiming the toxicity and carcinogenicity of synthetic pigments. However, any substance can have an adverse effect when a biased opinion is applied. The natural colorants, carotenoids and lycopene, have received attention as natural antioxidants. Flavonoids (anthocyanins) have been considered not only as natural antioxidants, but also as phytochemicals as the primary function of flavonoids is as plant hormone substances. The melting point of lycopene is 172 °C, and beta-carotene is 181 °C and substances with high melting points are difficult to metabolize or discharge from our bodies, but there is no reason to avoid using natural pigments in foods. Most people agree that the use of extracted or synthesized natural colorants is a good idea, however, their high melting points and the difficulty in metabolizing them proves otherwise. The hope that antioxidants can prolong our lifespan has disappeared after a study on resveratrol (an antioxidant in grapes and peanut roots). Utilizing more natural colorants requires many additional clinical studies and more regulations. In conclusion, there are no any ideal pigments in nature.

Perception varies with individual differences and conditions

Differences due to age and sex

Newborn babies have the most sensitive perception of taste and smell as they have taste buds throughout their entire mouth. They have taste receptors in their palate, throat, and the sides of their tongue. Owing to their great number of taste receptors, babies taste diluted powdered milk several times stronger than adults do. The extra taste buds disappear at around the age of 10, and after this the remaining taste buds repeat the process of cellular degeneration and production. From the

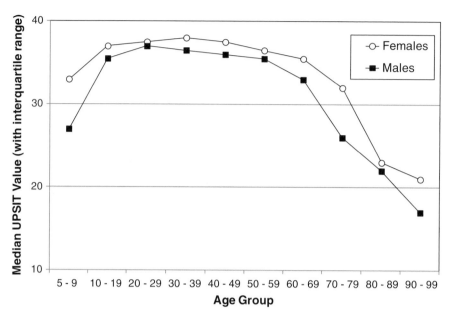

Figure 3.5 Ability to distinguish odors by age groups. Modified from Longo et al. (2011).

age of 20, the sense of smell becomes reduced, and after the age of 60, it drastically diminishes (Figure 3.5). At the age of 80, three out of four healthy seniors cannot perceive smells very well. The majority of the population between the age of 65 and 80 goes through a serious case of anosmia. It could be one of the reasons why seniors cannot enjoy their food as well as they used to.

The results of many experiments since 1899 have revealed that women are more sensitive to bitterness and odors than men. It was mainly only men who were responsible for hunting animals for food, while women were responsible for picking edible plants. The majority of toxins are produced by plants, not by animals. Thus this may be the reason why women are more sensitive to bitterness and odors. One interesting experiment revealed that men are more sensitive to bourgeonal than women. Professor Matthias Laska's team in Sweden conducted an experiment with 250 men and 250 women using the main component of bourgeonal, which is a fragrance reminiscent of lily-of-the-valley (Olsson and Laska, 2010). The results showed that the average threshold value of the odorant was 13 ppb for men and 26 ppb for women, which showed that men were twice as sensitive as women. The research by Dr Marc Spehr from Ruhr-Universität Bochum in Germany showed that human sperm can recognize and move towards bourgeonal (Spehr *et al.*, 2003). Since the sensory receptors that perceive bourgeonal exist not only in the olfactory epithelium in the nose but also in the cell membrane of sperm, curious scientists conducted this test to see if men would be more sensitive to the smell.

Individual variation is also significant

A team lead by Dr Paul Breslin in the Monell Chemical Senses Research Center revealed that each individual has different gustatory receptor genes that perceive bitterness, and, consequently, carries their own distinctive set of taste receptors, which gives them unique perceptions. His research team discovered that there are several types of *hTAS2R38* genes of bitterness receptors, and people with the most sensitive receptors are 100–1000 times more sensitive to the bitterness than the type of people with less sensitive receptors (Bufe *et al.*, 2005). It seems that everyone may have their own world of taste. The difference is greater for the olfactory sense than the gustatory sense. Out of the many genes, it is rare to see more variation of sensory genes between individuals that are not olfactory genes. There are 388 kinds of olfactory receptors, but anosmia was found for around 50 of them. Just as those with color-blindness or color-weakness see the world differently from regular people, the pattern of tastes that people perceive from the same food could also be very different; this proves why everyone may not have the same preference for a particular food. However, because we use the same descriptions, we may believe that we taste food the same.

The same smell of sweat can be felt differently by various individuals. A joint research team at Duke University and Rockefeller University found that individuals smelled sweat to be like urine, vanilla, or as nothing (Keller *et al.*, 2007). The researchers focused on androsterone and the olfactory receptor *OR7D4* genes. After collecting blood samples from 400 people, the differences in their genes were investigated. According to their investigation, some people perceive the smell of sweat or the smell of other people to be offensive (sweaty or urinous), pleasant (sweet or floral), or not at all depending on the differences in their genes. Their results verified the link between the function of a human odorant receptor and odor perception.

The research team of Dr Ivanka Savic at Karolinska University in Sweden revealed that the hypothalamus of women and nonhomosexual male-to-female transsexuals are activated by the hormones in men's sweat (Berglund *et al.*, 2008). When they were exposed to the odor of testosterone (male hormone), through positron emission tomography images they showed strong sexual reactions in the brain of the female and transsexual subjects. In contrast, when estrogen (female hormones) was tested, the subjects did not display any activity. Ordinary smells, such as that of lavender, all displayed normal activity in the part of the brain in charge of the sense of smell. However, testosterone activated the reactions only in women and transsexuals, while estrogen activated the reactions only in men.

There is a theory that the reason for using perfumes is to strengthen the individual body odor. This theory is an important clue in understanding the uniqueness of the perfume industry, which cannot be considered like other fashion industries, such as clothing, cosmetics, and accessories. The individual preference to certain scents is so different from one person to another, and does not simply follow fashion trends. Strictly speaking, trends do not exist in the perfume industry. There is not one type of perfume that everyone universally prefers. This statement can be confirmed by the fact that the present best selling perfume brands in the world have been around for more than 50 years. "Mitsuko" perfume from Guerlain was first released in 1919, "Chanel Number 5" was released in 1921, and "L'air du Temps" from Nina Ricci was released in 1948. In a word, people persistently prefer their own favorites.

According to a research article published by Dr Matha McClintock's team in *Nature Genetics*, the male body odor plays an important role in attracting the opposite sex (Jacob *et al.*, 2002). Each person has their own distinct body odor because of the substance known as MHC (major histocompatibility complex) that exists within our body's immunocyte. They found that women can detect differences of one human leukocyte antigen among male odor donors with different MHC genotypes. Their data show that HLA-associated odors influence odor preference and may serve as social cues.

Claus Wedekind, an evolutionary biologist in Sweden, and his colleagues proved the involvement of the MHC gene in human relation preference when women are exposed to certain body odors of men (Wedekind *et al.*, 1995). Each male student wore the same T-shirt for two days. The third day, each female student rated the odors of six T-shirts. They scored male body odors as more pleasant when their MHC differed from the men than when they were more similar. Furthermore, the odors of MHC-dissimilar men remind the test women more often of their own actual or former boyfriends than do the odors of MHC-similar men. This suggests that the MHC or linked genes influence human mate choice today. One possible explanation for this phenomenon of different MHC genotype preference would be a means of avoiding mating partners with closely related genes.

Fragrances have been used for at least 5000 years. However, the significance of large individual differences in preference for fragrances has been an unanswered question for just as long. It may be because the preference for various perfumes relates to the individual diversity of genes. To prove this hypothesis, evolutionary biologists Milinski and Wedekind (2001) asked 137 male and female college students to review 36 types of perfumes that were being sold on the market. Afterwards, they were asked two questions: "Would you like yourself to smell like that?" and "Would you like your partner to smell like that?" As expected, the preference to perfume for himself/herself could be easily classified according to their MHC genotypes. However, the preference for perfume for their partner did not match with their partner's MHC genotypes, which the test subjects did not know about. The results agree with the hypothesis that perfumes are selected "for self" to amplify in some way body odors that reveal a person's immunogenetics. Therefore, there is no one perfume that is preferred by everyone, because of the incredible diversity of MHC genotypes that determine the perfume preference of individuals.

Differences due to race and history

Non-Caucasians generally have more sensitive olfactory perception than Caucasians. Wherever we visit, we find that the most preferred food in the region is different from those in other areas. The hedonic preference to odors is greatly influenced by the culture and environment. This influence is a very unique characteristic of olfactory sense but no other sensory perceptions. In the United States, Japanese pyrola (*Pyrola japonica* Klenze) is very often used as a minty flavor for candy, and, thus, many Americans like this herb. However, in the United Kingdom, the same herb extract was used for painkillers, and, therefore, it is does not give off a pleasant feeling for the British. There are also many cases where some European certified perfumers could not distinguish between certain traditional Korean aromas, which can be easily recognized by Korean certified perfumers.

- Common discomfort odors: smell of fish, excrement, body odor, putrefaction, and mercaptan (natural and propane gas additive).
- Common pleasurable odors: smell of plants, lavender, amyl acetate (banana flavor), galaxolide (musk flavor), and eugenol (clove flavor).

- Cultural difference:
  - Existence of secreting glands (Europeans and Africans, unlike Asians, have more apocrine sweat glands).
  - Differences in hygiene habits, cultural norms and manners.
  - Familiarity with specific odors: sesame oil is a symbol of savory aroma for Asians, but Westerners who have never used it before, perceive it differently; as Asians consume more beans for protein sources than milk, they are not familiar with the flavor of cheese.
  - Experience and imagination: due to the absence of cremation in their culture, East Asians cannot understand why Westerners dislike the grilling smell of fish jerky.

The preference for smells is constantly changing

When the majority of Americans lived in farming areas, the smell of livestock meant a decent and comfortable family life. Nowadays, the same smell offers a different feeling for urban dwellers visiting farming areas. They regard the smell of livestock as air pollution and also dislike the smell of manure for agriculture. They like organically grown fruits and vegetables, but promptly close their car windows when they smell manure on the freeway. The preferences towards smells change over time. Nowadays, there are not many people in urban areas who are familiar with the smell of burning leaves. In ways such as these, there are no senses that are absolutely preferred. Everything is always changing.

In reality, these are only some of many examples. Enjoying a dinner with loved ones in a high class restaurant with a luxurious decor, sweet music, great customized service, and fabulously decorated food, is completely different from eating the same food in a dark dungeon filled with screams after being kidnapped by a rugged looking man. Actually, even eating the same food in the same environment but without light and in the dark may provide a completely different taste. Saying that the taste and smell take up the majority of the flavor of a food is only applicable when all other conditions are the same. In fact, the taste and smell of the food are considerably influenced by the psychological conditions. The principle of taste and smell is inside our brain. Therefore, in order to properly understand the perception of taste and smell, it is necessary to understand the recognition process in the brain for sensory perceptions.

References

Berglund, H., Lindström, P., Dhejne-Helmy, C., Savic, I. (2008) Male-to-female transsexuals show sex-atypical hypothalamus activation when smelling odorous steroids. *Cereb. Cortex.* **18**(8): 1900–1908.

Bufe, B., Breslin, P.A.S., Kuhn, C., Reed, D.R., Tharp, C.D., Slack, J.P., Kim, U-K., Drayna, D., Meyerhof, W. (2005) The molecular basis of individual differences in phenylthiocarbamide and propylthiouracil bitterness perception. *Curr. Biol.* **15**: 1–6.

Jacob, S., McClintock, M.K., Zelano, B., Ober, C. (2002) Paternally inherited HLA alleles are associated with women's choice of male odor. *Nat. Genet.* **30**: 175–179.

Keller, A., Zhuang, H., Chi, Q., Vosshall, L.B., Matsunami, H. (2007) Genetic variation in a human odorant receptor alters odour perception. *Nature.* **449**: 468–472.

Longo, D.L., Fauci, A.S., Kasper, D.L., Hauser, S.L., Jameson, J.L., Loscalzo, J. (2011) *Harrison's Principles of Internal Medicine*, 18th edn. The McGraw-Hill Companies, Inc., New York.

Milinski, M., Wedekind, C. (2001) Evidence for MHC-correlated perfume preferences in humans. *Behav. Ecol.* **12**(2): 140–149.

Olsson, P., Laska, M. (2010) Human male superiority in olfactory sensitivity to the sperm attractant odorant Bourgeonal. *Chem. Senses.* **35**(5): 427–432.

Spehr, M., Gisselmann, G., Poplawski, A., Riffell, J.A., Wetzel, C.H., Zimmer, R.K., Hatt, H. (2003) Identification of a testicular odorant receptor mediating human sperm chemotaxis. *Science.* **299**: 2054–2058.

Taylor, A.J., Linforth, R.S.T., Baek, I., Martin, M., Davidson, M.J. (2000) In: *Frontiers of Flavor Science* (ed. P. Schieberle, K.-H. Engel). Deutsche Forschungsanstalt für Lebensmittelchemie, Freising.

Wedekind, C., Seebeck, T., Bettens, F., Paepke, A.J. (1995) MHC-dependent mate preferences in humans. *Proc. Biol. Sci.* **260**(1359): 245–249.

4 How do we Smell Odors?

How do we smell odors? The area for sensing olfactory stimuli is located on the upper part of the nasal cavity closest to the brain. This area uses only a portion of the air that is inhaled to detect odors. The olfactory membrane region is easily distinguished by its yellowish brown color. It is symmetrical and sized around $2-4\,cm^2$, about the size of a small coin. It is the place where around ten million nerve cells, called olfactory cells, are positioned. Right below the olfactory cells, there are countless cilia, which hold around 1000 olfactory receptors on their surfaces. There are many types of olfactory cells. If there were only one type, the olfactory cells would only be capable of distinguishing just one kind of odorant. Although all olfactory cells look identical, they can distinguish many types of odorants, and only recently it was discovered that each olfactory cell identifies different odorants. There are several types of theories behind the sensing mechanism and how cells are able to distinguish such a wide variety of odorants. Here we will discuss the most popular theory, the Amoore molecule structure model, which uses the characteristic shapes of molecules (Amoore *et al.*, 1964). This stereochemical theory hypothesizes that every odorant has a different molecular shape (circular, oval, bottle shape, *etc.*) that will fit into its respective pore of the olfactory cell to produce a neuro-signal specific to its odor (Figure 4.1).

Olfactory receptors are G-protein coupled receptors

Amoore's stereochemical theory revealed that odorants approach the surface of olfactory cells to activate "something," which results in a

How Flavor Works: The Science of Taste and Aroma, First Edition.
Nak-Eon Choi and Jung H. Han.
© 2015 John Wiley & Sons, Ltd. Published 2015 by John Wiley & Sons, Ltd.

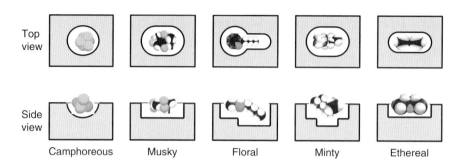

Figure 4.1   **Amoore's stereochemical theory.**

signal to the brain. The identity of this "something" was discovered to be a receptor in 1991. In 2004, Richard Axel and Linda Buck received the Nobel Prize for discovering that G-protein coupled receptors (GPCR) identify the odorants in the olfactory cells. The G-receptors are small compared with the size of other proteins. They are present in the cell membrane and penetrate back and forth through the membrane seven times. Therefore, 7TM (trans-membrane) receptors might be a more appropriate name for the GPCR. However, the concept of a signal transmission mechanism using G-proteins had already been discovered in 1994 and been awarded the Nobel Prize. The operational receptors, which were discovered by Linda Buck in 2004, were thus named G-protein coupled receptors (GPCR). The special function of the G-protein is to detect molecular shapes from outside the membrane (Figure 4.2). Presently, approximately 800 types of G-receptors have been discovered in the human body. Therefore, our body is capable of distinguishing and recognizing at least 800 types of molecules.

These 800 G-receptors have the same basic structure, but the three-dimensional conformation of the G proteins can change depending on their amino acid composition. The relationship between the G-protein and flavoring chemicals is like that of a lock and key. In other words, the shape of each keyhole (G-protein receptor) is different. Since the shape of the keyhole is different, the possible shape of the molecule that fits into the hole is also different. The G-receptors continue to move around the surface of the olfactory cells until the right shape of odorant (key) comes around. Then it binds with the molecule and changes its shape to an "ON" state. The only major function of G-receptors is to generate an "ON" signal after binding with the right odorant. The actual signal is formed by the G-proteins throughout a series of enzymatic reactions. The G-receptors play the essential role of identifying the specific odorant. Even though the

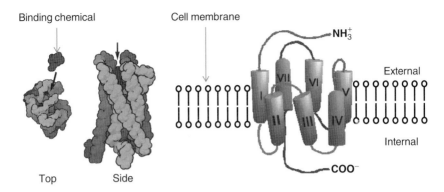

Figure 4.2   A typical image of G-receptors and the competitive image of molecules. Modified from Wikipedia Commons (http://en.wikipedia.org/wiki/File:Mu-opioid_ receptor_(GPCR).png) and (http://en.wikipedia.org/wiki/File:GPCR-Zyklus.png) (accessed 17 July 2014).

biological locking mechanism is sufficiently sophisticated, it is not perfectly accurate, as many more than 800 molecules can also "open the locks" (G-receptors). Although our naked eyes cannot perceive small molecules, our body can, however, always recognize them.

As stated earlier, the existence of umami, as has been asserted by Professor Ikeda since 1907, was completely and officially accepted after glutamic acid receptors in the tongue were found. These glutamic acid receptors are a type of G-receptor. Sweetness receptors are also G-receptors. Therefore, it is necessary to understand G-receptors in detail to fully understand the mechanisms of sensory perceptions. It is not just the senses of smell and taste, but also the visual sense is perceived by G-receptors. Moreover, there are many G-receptors for recognizing hormones and neuro-transmitters. Not only do sensory perceptions depend on G-receptors but also survival does too. The hormones related to the feelings of satisfaction and obesity are mostly perceived and detected by G-receptors. Many effective medicines for immunity and cancer treatment are also related to G-receptors. Understanding G-receptors will help to comprehend the principles of other receptors because they all follow a similar principle.

G-Receptors differentiate isomers, resulting in different odors

Many substances with the same molecular formula and molecular weight do not have perfectly identical three-dimensional structures due to isomerism, such as geometric isomers and stereoisomers.

Because of the different possible conformations of an odorant, that is, its isomers, the odorant may not always bind with the same G-receptors and, consequently, also yields a different smell. There are two common types of geometrical isomers for the molecules with double bonds (unsaturated bonds): the chair shaped trans isomer and the boat shaped cis isomer. This means that the molecular conformations are definitely different. Stereoisomers are harder to understand. Their difference is analogous to the symmetric (mirror imaged) conformations of a left-hand and right-hand glove. The way to understand isomerism is to realize that each isomer of an odorant can be distinguished by the olfactory sense. If there are substances with the same molecular weight but different structures, which substance would be recognized easier based on its smell? The answer is, of course, the isomer that has existed in nature for the longest time. Our sensory perceptions have evolved for the purpose of sensing substances that exist in nature. They have not developed, or advanced in a way to sense substances that do not yet exist. Sensory organs feel only the substances that we need to recognize out of all the molecules in nature.

For geometric isomers, there are more cis forms than trans forms in plants in nature. Therefore, cis isomers have a stronger odor. With stereoisomers, the more common form in nature has a stronger smell. Menthol has eight stereoisomers due to its three chiral carbons. However, the most common form in nature is L-menthol, which gives off a continuous refreshing smell. L-Menthol has the strongest odor compared with other forms of stereoisomers. For example, nootkatone is an odorant of grapefruits. It also has eight stereoisomers, but the natural nootkatone from grapefruit has the strongest smell, while the other rarer isomers give a weaker taste and odor. Geosmin is the earthy-odor chemical from beets. Like other natural odorants, the natural form, (−)-geosmin, has an intensity that is 11 times stronger than the synthetic form, (+)-geosmin. Among a variety of isomenthols, neo-isomenthol has the strongest smell, the intensity of which is five times that of other menthols. For a rosy smell, the natural form, rose oxide, has an intensity 100 times stronger than that of other synthetic rose odorants. Thus the smell of natural substances is clearly stronger than their synthetic isomers.

When odorants are produced by enzymes through a natural biochemical pathway, the final odorants are in the form of just one of their stereoisomers. However, when they are synthesized using chemical catalysts through organic chemistry, several stereoisomers are produced. Generally, these isomers have a weaker smell. The sensory organs in our bodies can identify the difference between the "left-hand glove" and the "right-hand glove." They can match the "hand" with the "correct glove." However, catalytic synthesis simultaneously creates both left-hand and right-hand gloves, which slows down the process of recognition of one specific glove to the exact hand due to the existence of the other useless glove.

If there were synthetic odorants that have stereoisomers perfectly identical to those of natural odorants, there would be no differences between their smells. However, as long as there is a stereoisomer, the intensity of synthetic odorants has to be weaker than that of natural odorants. The reason natural products are generally thought to be weaker with respect to their smell compared with synthetic products is that natural odorants are sold on the market with very low concentrations and at expensive prices. The synthetic odorants are generally used at higher concentrations in commercial products with inexpensive prices. Regardless of the natural and synthetic odorants, the main chemicals of a specific odor are the same substances. Only the source and price of the odorous chemicals are different. Synthetic odorants are as strong as natural ones, but they are added to commercial food products and perfumes more than the natural odorants are.

G-Receptors perceive multiple chemical substances

G-receptors utilized in the perception of smell are located in the olfactory cavity (known as OR and TAAR) and the vomeronasal organ (known as V1R and V2R), an auxiliary olfactory sense organ. For the perception of taste, there are taste receptors (T1R and T2R). The human tongue has approximately 10 000 taste buds, dogs and cats have less, but pigs have more than 10 000 taste, while cows have around 30 000. Pigs and cows have to use their tongue to instantly judge the taste of food inside their mouth because they eat many different types of plants. However, dogs and cat have less need to sense taste because they hunt other animals to eat. The taste buds have multiple types of taste receptors, and one taste receptor can sense one kind of taste. So far, 40 types of G-receptors (T1R and T2R) responsible for the perception of taste have been discovered. Out of these receptors, a large number (T2R) account for bitterness and of the 36 kinds of bitterness receptors (T2R), 11 of them have been fossilized, and 25 are active (Figure 4.3). Since there are many types of toxins that are bitter in nature, it is very important to be able to recognize them. Sour and salty tastes are sensed by channel receptors rather than G-receptors, and sweet and umami tastes are sensed by

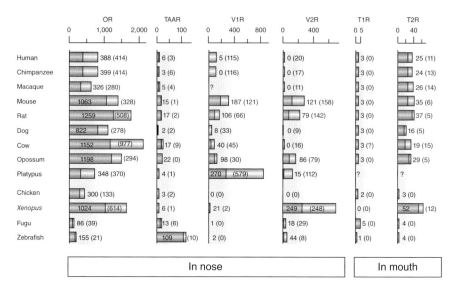

Figure 4.3 Comparison of the taste and olfactory receptors in various animals. Values in parentheses are the number of fossilized receptors. Modified from Nei *et al.* (2008). (See color figure in color plate section).

T1R. But what is most surprising is that there are only three kinds of T1R: one for sweetness, and two for umami. This is observed in the majority of other animals as well. The mostly preferred taste, that is, sweetness, is recognized by only one receptor regardless of whether they are natural or synthetic sweeteners. Therefore, the human body perceives every possible taste with one receptor for sweetness, two for umami, one for saltiness, one for sourness, and 25 for bitterness.

The receptors for smell (OR) and pheromones (TAAR, V1R, and V2R) are all G-receptors that exist in the nose. The human nose has more than 388 kinds of G-receptors. Originally there were over 800 but the majority of them have degenerated. The estimated number of G-receptors varies depending on the researchers. The numbers range from 375 to 420, but in this book we will assume 388 kinds for convenience. The number of G-receptors is pretty large. When we consider this one function (the perception of smell) compared with the multiple different functions of other organs, the number of genes for 388 receptors may be the largest for just only one function. Our body has just over 20 000 genes, and although our visual sense is responsible for 90% of all information collected, it requires only three genes, whereas a much greater number of genes are incorporated into the perception of smell. At present, the human perception of smell has degenerated significantly. However, it is associated with multiple genes and is called the sensory of instinct, because it has much more influence on our brain function than we typically think. However, overall, most pheromone receptors have degenerated. From Figure 4.3, it is easy to understand how much the perception of smell dominates the behavior of animals whose pheromone receptors are still active. We will discuss the hidden functions of the olfactory sense in later chapters.

G-Receptors work simply as on/off switches

The G-receptors that are located on the cell membrane are normally in an "off" status until they bind with their specific molecules and change their structures to switch into an "on" status. In this state, the binding of the G-protein creates a secondary signaling substance that begins the signal transduction. This is the sole function of G-receptors, there are no other functions. However, because this simple function is connected to signal transduction mechanisms, the G-protein is involved in numerous other activities. This signaling process is connected to the brain to trigger the cognitive ability. The only function of olfactory cells is to deliver the resulting signal of the G-receptors to the connected nerve cells, which will transfer that signal to the brain. The mouth senses taste, the nose senses smell, and the eyes sense light. However, there is little difference between the functions of the sensory cells, besides their locations. The sensorial function of the sensing organs is simply a transmission of electrical signals to the brain, it is not the transmission of the actual taste or smell that is delivered. Electrically, sensing is the function of switching on and off. Thus, it is very similar to binary computer codes made up of 1s and 0s. Just as the data processed in a computer follow the rules of binary codes to create numbers, letters, pictures, music, and videos, the sensory response of G-receptors (i.e., data) are processed by the brain using the rules from the signals. When glutamic acid is in the mouth, it triggers the perception of umami, although if it is in the brain, it becomes a neurotransmitter.

However, it is not just sensory cells that can sense perception. Every other cell has the ability to perceive environmental changes as well. There is no reason for our body cells not to have the ability to sense when we consider that small bacteria, which are less than 1/1000 of one of our body cells, have the ability to sense environmental changes. The only difference between other body cells and sensory cells is that body cells do not deliver the sensory signals to the nerve cells. Instead, they use the signal transduction within their cells to establish when to control certain functions and maintain homeostasis. The intercellular control system can change the function of the target proteins or initiate the transcription of target protein DNA. The role of sensory cells is completed once the signal is delivered to the brain. However, body cells have to respond to the signal transduction in order to control the amount of target protein or to alter the function of the protein. The most common example is the control of the body's metabolism with respect to hormones. Hormones bind with the G-receptors on the surface of the cell membrane, and then the secondary signaling substances from the receptors trigger the linked DNA transcription, resulting in the production of a significant number of target proteins (i.e., enzymes). G-receptors are merely on/off switches located on the membranes of both sensory cells and also other cells.

Depending on the binding affinity to receptors, similar molecules can be recognized as completely different tastes and odors

The G-receptor is a switch to turn on the signal transduction with the correct key (chemical substance). However, it does not keep the switch in the on-position long term. It only stays on while the key is bound. All molecules move around and vibrate energetically. In this way, the molecules pair up with the right G-receptors to turn on the switch, and, in the same way, leave the receptors turning them back off. The duration of the binding time with the receptors varies considerably depending on the binding affinity of the molecules to the receptors. Furthermore, the extent of signal amplification varies depending on the concentrations of enzymes participating in the signal transduction. The capacity for amplification and cancellation also changes during the selection and integration processes of nerve cells. Finally, the sensitivity of the signal recognition varies from 1 to 1 000 000 times the magnitude, according to the differences in binding affinity, extent of amplification, signal integration, noise elimination, and contrast enhancement.

The binding affinity of chemicals to the receptors is affected mainly by electrostatic interaction. The binding force is similar to the attraction of magnets, as it is only effective when they are close, their charged parts are well aligned towards each other, and the geometrical molecular conformation is suitable. The stronger the interaction, then the stronger binding force that results. A stronger binding force means it is harder to separate the key from the keyhole, resulting in a longer binding time. All molecules vibrate, and consequently, dissociate from one another. A molecule will move in all directions and collide with neighboring molecules. The molecular movements are not strong enough to break the covalent bonds of the molecules, but they are strong enough to separate the electrostatic interactions between the other molecules. Therefore, they continue the process of association and dissociation within a short time frame. A strong binding affinity lowers the frequency of dissociation and also gives off a stronger intensity of the sensory perception. However, an excessively strong binding affinity has an adverse effect on ordinary sensory perception as it becomes extremely difficult to separate the bound substances. As an example, biotin, one of the B vitamins, fits perfectly with avidin, a protein found in high concentrations in egg white. Their binding force is so strong that the predicted period of separation of biotin from avidin at room temperature is 19 years, on average. Such a strong binding force is unsuitable for signaling transduction. If biotin was a sweet substance and avidin a receptor on the tongue, the sweet taste from a small amount of biotin would last for 19 years.

There is a strong synergism in umami when amino acid types (MSG) and nucleic acid types (IMP or GMP) are blended, which means that compared with using just MSG or IMG alone, when mixed together the intensity of umami increases. This synergism is caused by the specific structure of the umami receptors. These are C-type, which have two G-receptors, while other taste receptors are A-type with one G-receptor. When nucleic acid binds to the upper of the two binding sites, the conformation of the lower binding site is altered and will have a much stronger affinity for MSG. Owing to this increased binding force of the lower binding site to MSG, after the upper binding site has been combined with IMP or GMP, the intensity of umami from the MSG becomes stronger.

The broad spectrum of the olfactory sense

Odorous chemicals are detected by the G-receptors on the surface of olfactory cells in the epitholium area. The sensing signals generated by the G-receptors and enzymatic pathways of the olfactory cells are transduced to the neurons (nerve ends). Multiple neurons are combined and form a spherical structure known as an olfactory glomerulus (Figure 4.4). The olfactory glomerulus is positioned on the surface of one of two olfactory bulbs. The selected and integrated signals from olfactory glomeruli are transmitted through the fibrous

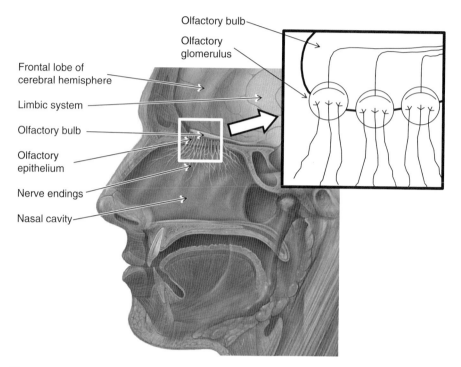

Olfactory bulb

Olfactory glomerulus

Frontal lobe of cerebral hemisphere

Limbic system

Olfactory bulb

Olfactory epithelium

Nerve endings

Nasal cavity

Figure 4.4   The region of the nose that perceives the sense of smell. Modified from Wikipedia Commons (http://commons.wikimedia.org/wiki/File:Head_olfactory_ nerve.jpg) (accessed 17 July 2014).

nerve cells in the olfactory bulb to the olfactory cortex of the brain. The glomeruli conduct the first level of synaptic processing, they do not simply deliver the signals from the olfactory cells to the brain, but first integrate all the signals from the multiple nerve cells. An fMRI system can visualize the activated functions of the glomerulus and can identify that a sensory substance stimulates several glomeruli. Just as only three visual receptors can react to a wide spectrum of light wavelengths, a broad range of olfactory receptors can be stimulated by one odorant. There are some differences in the degree of reactivity of receptors to one stimulant and in the binding affinity of the receptors to the substance. The fact that the signal integration function of multiple glomeruli for one odor is very complicated can be understood when we take into account the multiple different signals from the receptors for one odorant and their various signal intensities. The glomeruli collect signals and integrate them through various control mechanisms, such as a feedback loop control. The multiple responses of glomeruli are also explained by the existence of

Table 4.1   Different Perceptions of Odorants with Respect to their Concentration

Odorants	When 100% pure	When diluted
Undecalactone	Odor of petroleum	Peach scent
Dimethyl sulfide	Odor of fish stew or dried laver (sea weed)	From 0.1%, smell of cooked foods such as cooked strawberry jam, condensed milk
Indole	Unpleasant odor of feces	From 0.0001%, floral scents like jasmine or gardenia
Furfuryl mercaptan	Extremely unpleasant odor	From 0.001%, smell of roasted nuts or coffee beans
Decanal	Strong odor of petroleum	From 0.1%, scent of oranges

numerous chemical groups (i.e., osomophores), so they are responsive to the odor for molecules with their molecular structure.

Indole has an unpleasant odor at high concentrations, but it is a major odor of jasmine at diluted concentrations. The spectrum of odor can be used to explain the phenomenon of different odors of one chemical at a range of concentrations, such as for indole. The phenomenon of the change in the spectrum of odor at various concentrations can be interpreted by the theory of two (or more) osmophores in an odorant. Two osmophores will have different binding affinities to their receptors. When the odorant has a higher affinity to receptor 1 than receptor 2, it will mostly bind with receptor 1, especially when the concentration is low. Thus, the respective floral aroma will be recognized by the brain from the main signal transduced by receptor 1. However, when the concentration of the odorant increases, it will also bind with the receptor 2. Specifically, after receptor 1 is saturated by the high concentration of odorant, receptor 2 will bind with the odorant more and more, which will cause the brain to perceive a completely different odor than in the case of a low concentration. Table 4.1 lists typical odorants that are perceived differently with respect to their concentrations.

The transduction of sensory signals

When G-receptors initiate a signal, signal amplification occurs in the sensory cell. The extent of the amplification depends on the concentrations of the enzymes participating in the signal transduction pathway. If the concentrations of the enzymes are lower than normal, the signal becomes weak. When the concentrations of the enzymes increase, the signal is amplified. This is the same principle as when our eyes take a minute to adjust to the light when we walk into a dark theater or out into a bright area. The signal transduction process through the nerve cells to the brain is not simple. The signals are transfered through a complicated interaction of neurons and several channels of feedback, diffusion, and convergence. In the case of the olfactory sensory system, it is the glomerulus that primarily integrates the signals. Rabbits have 50 million olfactory cells and 2000 glomeruli, and dogs have over 100 million olfactory cells, but just several thousand glomeruli. The 10 million human olfactory cells are integrated with several thousands of glomeruli. Since an average of 25 000 olfactory cells combine into just one glomerulus, there is no larger scale of signal integration than the glomerulus and the sense of smell.

If every single sensory cell were individually connected to the brain, it would overload the brain's capacity. The brain functions more efficiently when there are fewer loads and more noise signals are filtered out, in which case the more useful signals will be clearly processed. Single cell organisms such as *E. coli* also have noise rejection circuits. If every single signal was processed, the meaningful signal would be easily lost. This ability to differentiate between useful and useless signals and to delete the useless signal is a critical ability of the brain. The most fundamental technique in signal processing is contrasted amplification. The main signal is amplified more when the surrounding signal is reduced. Through this contrasted amplification process, the strong signals are amplified, while the weak signals are diminished, which allows for a sharper contrast between the signals. The fundamental goal of signal processing is to focus on collecting meaningful signals by ignoring the ordinary signals and amplifying the contrast. Furthermore, the sensory system not only has the general signal transduction, but also a feedback system that receives the control signals from the brain. When we are hungry, our olfactory sense becomes more sensitive; however, it becomes less sensitive when we are satiated. Thus, instead of transmitting signals in just one direction from the sensory cells to the brain, there are also circuits that control the activities of the sensory cells through the control signals from the brain.

Olfactory fatigue is also a functional activity for life

When we are continuously exposed to the same odor, our brain loses interest in and the sensitivity of our olfactory perception to this odor is decreased. This olfactory fatigue is the best method for rejecting unpleasant odors and finding new odors. The sensitivity towards a continuous odor declines, in general, at 2.5% per second, and nearly

70% of the sensitivity is lost in 1 minute. Our olfactory sense can easily become unresponsive to any unpleasant odors. In addition, it is not usual to respond to a specific odor for a long time, and even the pleasant aroma of perfumes or herbs cannot be recognized after a long exposure. This phenomenon is called "olfactory fatigue (or adaptation)." However, during this state of exhaustion, other odors are easily distinguishable, in fact the sensitivity to other odors is enhanced. This phenomenon is called selective adaptation. After visiting a specific place frequently, the sensitivity to the smell of the place becomes decreased, which is called habitual adaptation.

We live in an environment surrounded by many odorants. It is impossible to make a perfectly odor-free room and everyone has a unique body odor. In reality, the odor-free state that we feel is the result of olfactory fatigue from environmental odors and our own body odors after staying in a specific environment for a long time. The reason the human olfactory sense easily becomes insensitive to a specific odor is to sustain its sensitivity to new odorants. The olfactory sense of nocturnal animals, especially to the body odor of predators, is far more superior than the sensitivities of their vision or hearing, and is of greater benefit for detecting predators at night. Moreover, the approaching predators can hide from the prey's visual and hearing senses, but it is impossible to conceal their odor. Also, all the other senses are ineffective during sleep, but the sense of smell remains active. So is olfactory fatigue (i.e., selective adaptation) the same sort of rest that muscles have after excessive exercise? Most certainly not: it is like the resetting of the baseline in gas chromatography with a control sample to maintain the sensitivity to the target peaks.

Surprisingly, the glomeruli of olfactory cells have a system of signals that also come from the brain. The brain directly controls the transduction of olfactory signals from the olfactory cells to the brain at the selection and integration stages. The lifespan of the olfactory cells is about 60 days. Therefore, there is no way that this number of cells can suddenly change their receptor structures or signal transduction pathways to cause olfactory fatigue. It is also unlikely that the body changes the concentration of signal transduction enzymes instantly just for the purpose of olfactory fatigue, but it is directly controlled by the brain.

Pheromones are designed to provide extreme pleasure at just trace amounts. When a butterfly identifies the smell of its pheromone from 4 km away, it will start to enjoy the pheromone. If the butterfly relocates just 1 km closer to the origin of the pheromone, the concentration increases 100-fold, and, consequently, the satisfaction of the butterfly caused by the pheromone will increase. However, if there were no olfactory fatigue phenomenon in the sensory system of the butterfly, there would be no desire for the butterfly to move closer to the origin of the pheromone because the concentration at 4 km away is already above the threshold of the pheromone effect. If the senses did not effectively become insensitive and merely continued to provide the same level of pleasure for a long time, the animal would refuse to do anything but stay where it was to enjoy the pheromone. Because of the strong effect of olfactory fatigue, the satisfaction from the specific smell decreases even for the same level of concentration over time. The animal will then search for higher concentrations of the smell to receive higher levels of satisfaction until it reaches the origin. Both the preservation and degradation of sensory sensitivity are vital activities for animals.

The recognition and integration of perceptions

The human body has over 388 types of olfactory receptors. There are several ways to figure out the brain's responses to the signals from specific and stimulated receptors. One of the most effective methods is functional magnetic resonance imaging (fMRI). This can observe the patterns of the brain's activity under the influence of a smell in real time. For example, the odor of almonds and the odor of yellow bananas stimulate different parts of the brain of the fruit fly. In this way, a map of the brain corresponding to individual odorants can be obtained.

In order to understand how the brain perceives an odor using an fMRI map, it is first necessary to understand the basics of how the brain functions. In a simple structural model of the parts of the mammalian brains, the primary targets for olfactory sense are the hippocampus, amygdala, corpus striatum, and thalamus. These four structures are all connected with the cortex and interact with each other in order to control actions and reactions. There is, obviously, a lot more involved with the working processes of the brain, however, we will only discuss the most fundamental functions and mechanisms of the brain that are relevant to the olfactory sense.

Parts of the brain

Hippocampus: The system involved in the olfactory sense sends most of its output signals to the hippocampus. The hippocampus was formed in the brains of small mammals as a structure to manage high-dimensional olfactory information and is a vital organ in charge of memory. If the hippocampus is damaged, new memories cannot be stored. The memories before the damage remain wholly intact but new memories cannot be formed. On this basis, it is hypothesized that the hippocampus is a temporary memory storage area before the information is permanently stored in the cortex. Without the hippocampus, new thoughts cannot be either perceived or stored.

Amygdala: The amygdala induces powerful emotional responses. When simple electromagnetic waves are sent to the amygdala via electrodes, the electrical stimulation can cause intense anger, dedicated love, depressing sadness, and almost every other psychological state man can experience. It also sends a large number of signals to the hypothalamus, which controls the endocrine glands. The secretion of hormones from these glands causes instinctive behaviors. Thus, the amygdala is the organ that specifically controls fundamental behaviors such as sex.

Corpus striatum: The striatum receives input from the cerebral cortex. It is also known for its function in the planning and modulation of movement pathways, but is also potentially involved in a variety of other cognitive processes. For higher-level animals, there are more variations in the area of motor output.

Thalamo-cortical loop: This is the loop that connects the frontal cortex with the thalamus. The link between the thalamus and cortex started off very small but developed into a very important region as the brain grew bigger in size. For large mammals, the thalamo-cortical loop is responsible for the core function of the brain. The olfactory cortex is connected to a specific thalamo-cortical loop. As the frontal lobe began to control the corpus striatum, it became able to decide what type of activity to be selected for a given situation. As a result, the ability to plan became possible.

Continuous circulations in the loop

Just like the olfactory cortex, the corpus striatum connects with the thalamo-cortical loop, and forms a bigger loop called a cortico-striatal-thalamo-cortical circuit. As the amygdala and hippocampus connect with the corpus striatum, it creates a bigger feedback circuit containing multiple loops. Therefore, olfactory information goes not only directly to the olfactory cortex but also through the corpus striatum, amygdala, or hippocampus. In fact, the olfactory sense is the only sense that is directly linked with the hippocampus and amygdala. Therefore, it has the unique feature of stimulating emotions and memories.

The link from the olfactory system to the amygdala causes emotional and natural responses such as thirstiness, satiety, pleasure, and doziness. From the amygdala, it connects again with the corpus striatum. The corpus striatum connects to the frontal thalamo-cortical loop where something surprising happens. The satiety hormone signals from the amygdala and hippocampus can be combined with the olfactory signals in the loop, resulting in an alteration of the final perception. In other words, the smell of food can be changed by this integration, so it can be perceived differently before and after eating.

The loops ultimately create intention and motivation, but they also cause sensory ambiguity following the integration and looping processes. Digital signals in computers and communication devices have their identification codes in the signal fragments. However, the signals of nervous systems do not possess the identification codes in their signals. When extra salt makes a food tastier, it is hard for consumers to differentiate whether the main change is caused by an increase in saltiness or an increase in flavor. The source of the taste signals becomes vague as the sensory perceptions are mixed during the signal processing in the loops. The integration of sensory signals is complex and accurate, but it is also not perfect. It is not precisely accurate due to various interferences that can result in many mistakes. However, it still provides sufficient reliability, and its inaccuracy also provides an opportunity for flexible adaptations.

Moreover, it explains the reason why the correct integration of all the senses is critical to create the right taste. Thus if a fruit does not taste sweet, the smell of the fruit feels empty and the lack of salt prevents all the other tastes from being recognized correctly, it is because of the integration of the sensory signals. For example, when one has very bad grades in school, the overall GPA (grade point average) is low despite having one or two As. However, when one has very good grades, their overall GPA is high despite having one or two Ds. Senses are the same. When a critical sense is not recognized, the whole taste and flavor of a food feels empty. But the most important thing is that the pleasure is also related to the looped integration system. The simple perception of senses has no meaning. Pleasure has to be rewarded when a certain action is advantageous for survival so that such actions will be conducted more often in the future. On the other hand, a pain (or displeasure) is a penalty used when the action is disadvantageous for survival so that such an action will be restricted later. The compensation system through pleasure is a critical element for survival that has developed and operated through the entire arrangement of the sensory recognition circuit.

G-Receptors can perceive light

The role of G-receptors is to detect molecules. A photon is far smaller than any molecule so it may sound ridiculous to state that G-receptors can sense light. However, it is true. Light can be perceived by G-receptors. The light receptor in the eyes is a rhodopsin protein, which is an opsin protein combined with retinol that is derived from vitamin A. This opsin is a G-receptor. A retinol molecule is imbedded in the middle of the G-receptors and forms a rhodopsin molecule. The G-receptors, sensing the molecules, can react with the light because of the photoisomerization activity of retinol. Retinol is in the form of cis-retinol in the dark, but when excited by light, it changes into a trans-retinol. This geometrical isomerization is the switch for the on- and off-state of the G-receptors.

The subsequent process of sending signals from a rhodopsin molecule is the same as that for any other receptor. Even with only weak signals from a few rhodopsin molecules, the amplification process allows us to perceive light anyway. The reaction of just one rhodopsin molecule can activate several hundred enzymes, and in a second, the enzymatic reactions can create several thousand secondary signal messengers known as cGMP (cyclic guanosine monophosphate). Thus, it is possible to visually perceive light with a very weak light source. Since the retinol is a substance derived from vitamin A, an insufficient supply of vitamin A can cause a lack of retinol, which is required for the functioning of the rod shaped cells in the retina. In such a case, when the number of operational G-receptors (with retinol) decreases, the eye can no longer detect weak light in the dark, leading, to night blindness.

During a long-term development process, such as adaptation, in evolutional biology, because the use of G-receptors for the sense of smell proved to be successful, they were also used for the visual sense, by gradually improving the efficiency of the G-receptors.

Understanding G-receptors can provide many answers

Around 800 G-receptors have been discovered in the human body. Nearly half of these are used for the sense of smell, around 30 are used for the sense of taste, and 3 are used for the sense of sight. But what are the rest of the receptors (just less than 400) used for? They are used for neurotransmissions, namely for the control of body physiology.

Sensory receptors send signals to the brain, but the other G-receptors play the role of controlling the vital functions through the development of genetic information, management of protein functions, and other activities. G-receptors can also detect noradrenaline, dopamine, histamine, acetylcholine, glutamic acid, and GABA (γ-aminobutylic acid) along with concentrations of peptides, proteins, lipids, organic matter, calcium, water, cell population, and more. These functions are involved, for example, in the division of cells, growth, movement, specific development, and death. The G-receptors are also associated with the maintenance of homeostasis and even cancer development.

In 2012, Robert J. Lefkowitz at Duke University and Brian K. Kobilka at Stanford University received the Nobel Prize for their distinguished research results on G-receptors. Since 1968, Lefkowitz's efforts to understand receptors within cells had resulted in the discovery of several types of hormone receptors, including adrenalin. After collaborating with Professor Kobilka and his team, they found that there are many more kinds of protein receptors with similar structures and they defined the protein structures on which the research fundamentals on G-receptors were built.

The press report on their results said, "The G-receptors are responsible for the important roles of reproduction, immunity, motorization, digestion, respiration, blood circulation, sleep, and others. Since G-receptors are related to various illnesses, they are especially important in the fields of medicine and pharmacy. Around 25% of the 200 biggest selling drugs to treat illnesses related to the central nervous system, cardiovascular system, inflammations, and metabolic syndromes, are products associated with G-receptors. Over 40% of the new drugs that are currently being developed are associated with G-receptors." The medical value of G-receptors has been well recognized, but this press release did not focus on the fact that the G-receptors are also responsible for the perceptions of light, taste, and smell that dominate our life.

In 2004, Linda Buck received the Nobel Prize for discovering that G-receptors are responsible for the sense of smell. Only eight years later, two Nobel Prizes were awarded to research teams who discovered the G-receptor. Most G-receptor research has focused on new drug developments. Around 50 G-receptors were subjected to such research, while there are still 100 G-receptors whose structures and activities are as yet unknown. However, G-receptors are nevertheless not popular research subjects in the biomedical area. The effective substances that can function as either medicines or toxins do not have any functionality in their structures. The medicinal effectiveness is only a switching function of the substances to turn the receptors on and off correctly. Any materials that have medicinal effects are not permitted as food ingredients. However, exaggerated claims of "astonishing health effects" are over-used in the nutritional supplement market. The G-receptors that actually control these "astonishing effects" do not receive much attention. The value of the Nobel Prize awards has, unfortunately, not been considered in a fair manner. G-receptors have various functions in:

- *Perception of cell density/population*: cancer and growth hormone control
- *Brain neurotransmitter*: behavior and emotion control
- *Foreign substance recognition*: immunity and infection response control
- *Autonomic nervous system*: blood pressure, pulse, digestive system control
- *Homeostasis*: calcium and moisture control.

Understanding G-receptors provides a big clue to the understanding of the true methods of regulating the effectiveness of poisons and medicines. The small amounts of regulating substances (i.e., signaling substances) that turn countless biochemical functions on and off in our body can result in many big changes. These signaling substances that are produced by our bodies are made up of carbon, hydrogen, oxygen, and nitrogen. It is important to create these substances in a form that is not too big and complicated, but that is still different from other organic chemicals. However, it is too difficult to produce them in a way that they perfectly unique compared with other organic molecules. Therefore, their shapes must be just sufficiently unique to be able to perform their roles well. Neurotransmitters are common organic substances. Hence, there is a chance that similar molecules exist in nature. Small amounts of these similar substances can act as medicines, while excessive amounts can be poisonous.

Pain and pleasure are illusions produced by the brain. For survival, our body creates a circuit of pleasure by using an endogenous morphine (endorphin) whose structure is similar to natural narcotics. Whenever our body acts in a favorable way to survival or reproduction, it rewards itself with pleasure. Narcotics can be described as any chemicals that have a similar form to endorphin or function in a similar way to endorphin. The regulated, low level of endorphin produced by the body provides pleasures rather than hallucinations. However, any foreign narcotics injected at a concentration beyond the simple pleasure level create hallucinations. No matter what type of molecules they are, if they can bind with the endorphin receptors, they create pleasure. The drug itself is not addictive, but the pleasurable hallucination of the brain is. Unbearable pain is also another illusion created by the brain. Ironically, there are some situations where the pain is so unbearable that some people would prefer death. However, this pain is actually an illusion created by the brain. When the sensory cells of pain are stimulated, they send their signals to the brain. Pain is the result of what our brain perceives from the signals. Injuries are disadvantageous for survival. When the injured part of the body delivers pain signals, the brain creates pain as a form of punishment so that such injuries may not be repeated frequently in the future. Thus, pain and pleasure are illusions created by the brain for the sake of survival. However, nobody is free from such illusions.

In fact, there are things called "death receptors." When an old cell is close to the end of its lifecycle, it has to be decomposed very quickly so that the next generation cell can be substituted without a gap. When the death signal is delivered to the cell, a large amount of an enzyme called caspase is produced through a complicated metabolic process inside the cell. The extremely high level of this enzyme is amplified instantly and decomposes the cell. This instant decomposition of the cell is caused by the intercellular system, and is not due to the high toxicity of the enzyme.

In reality, the mysterious functions of body physiology are the result of the control systems, and are not due to any biochemical substance. The substances are merely acting as keys that activate or inactivate the switches. Substances that have medicinal effects are essentially not allowed as food ingredients. Therefore, no food products can exhibit any great medicinal or toxic effects. They serve as supplements to their deficiencies. The idea that some substances are good and others are bad is a type of totemism. The mystery of life lies in the control network, not in the chemical substances. Any mystery is part of the whole system. No matter what the substances are, they become constituents of the system or keys to the switches. There is no mysterious power in their chemical structures. Molecules have a specific shape, size, and movement. They do not think. Everything is simply determined through the powerful control of our body systems.

Pheromones are not mysterious substances

A pheromone is a chemical that is specifically acknowledged within a particular animal group. When a substance is created and then perceived by a specific animal, with others not normally noticing it, it causes behavior in an agreed manner in that group. The agreed behavior is generally related to reproduction. Butterflies travel several kilometers to find their mates. A group of fish can perceive the chemical signals from a few fish and travel several kilometers to reach the location. Pheromones are not common substances, and they do not contain anything special. It is the agreement that animals from the same species have made that is special.

There is nothing in nature that has been explicitly created for human beings. We simply learn to discover, feel, and use a small part of nature. Such useful functions are created through a long period of evolution and through endless trial and error. It is also very costly to maintain such functions. Therefore, only the most urgent signals for survival are perceived and inherited. Toxins and medicines are effective because our body has a system that responds to them. It is not because the substances themselves contain toxic or medical effectiveness in their chemical structures. Toxic and medicinal substances simply function as switches that operate our body's system. It is a misunderstanding to think that the substances themselves contain special powers or functions.

References

Amoore, J.E., Johnston, J.W., Martin, R. (1964) The stereochemical theory of odor. *Sci. Amer.* **210**(2): 42–49.

Nei, M., Niimura, Y., Nozawa, M. (2008) The evolution of animal chemosensory receptor gene repertoires: Roles of chance and necessity. *Nat. Rev. Genet.* **9**(December): 951.

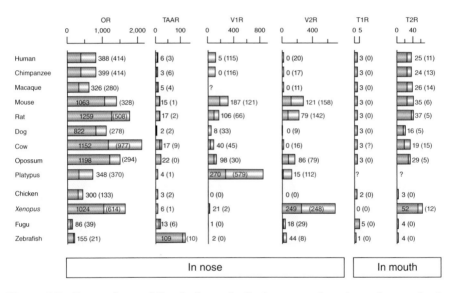

Figure 4.3 Comparison of the taste and olfactory receptors in various animals. Values in parentheses are the number of fossilized receptors. Modified from Nei *et al*. (2008).

How Flavor Works: The Science of Taste and Aroma, First Edition.
Nak-Eon Choi and Jung H. Han.
© 2015 John Wiley & Sons, Ltd. Published 2015 by John Wiley & Sons, Ltd.

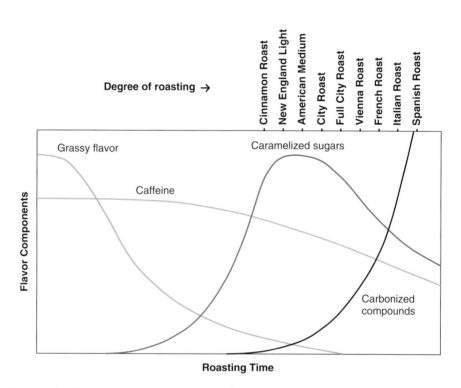

Figure 5.2 Heating decreases the grassy flavor of coffee beans while enhancing the color and other flavors. Modified from (www.therappahannockcoffee.com/roasting/) (accessed 17 July 2014).

5 What Creates Smell?

Odorous molecules are mainly created by plants

There are around 30 million different types of chemicals in the world. Among these, 95% are carbon-based compounds. Scientists synthesize around 2000 new chemical substances every year, which would add up to 200 000 new substances in 100 years. If people are responsible for only 200 000 types of substances, how are the remaining chemical compounds made? They are mainly synthesized by plants. People synthesize new compounds every year, but we do not use that many of them and only items that have any economic value are mass produced. As the rest are synthesized by plants naturally, it could be said that plants are the greatest chemists (or the biggest chemical plants).

The 30 million chemical compounds were not produced by chance. They were synthesized through several stages during the reaction processes of methylation, acylation, oxidation and reduction, formation of aromatic rings, and other complicated and concise organic chemistry reactions. Plants have many enzymes in order to carry out these biochemical reactions. Therefore, plants have many more genes than animals and human beings. The number of genes in potatoes or rice is far greater than that of their consumers.

However, we do not obtain all of the odorous chemicals from the 300 000 known plants. We obtain them from about 60 families of plants, which represents around 1500 species, and more than 90% of the odorants are obtained from less than 20 breeds of plant. The odorants are secondary metabolites, which are typically terpenes or phenolic compounds (Figure 5.1). Except for the "green note" of fresh cut grass or

How Flavor Works: The Science of Taste and Aroma, First Edition.
Nak-Eon Choi and Jung H. Han.
© 2015 John Wiley & Sons, Ltd. Published 2015 by John Wiley & Sons, Ltd.

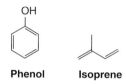

Figure 5.1   Chemical structures of phenol and isoprene.

vegetables, which is created from the oxidation of lipids, most odorant chemicals that plants produce are phenolics or terpenes.

Phenol has a hydroxyl group (—OH) on a six-carbon ring structure. A vast range of divergent types of phenolic compounds (also known as aromatic compounds) are produced when various atoms or other rings are added to the phenol structure. Anthocyanin pigments and lignin are substances that are created in the same way. If the terpenes provide the common flavors of herbs and spices, then phenolics provide the distinguishing flavor. The representative characteristic flavors of dried clove buds, cinnamon, anise, vanilla, thyme, oregano, and others, are phenolics. As a result of the hydroxyl (—OH) group on the phenol rings, phenolic compounds dissolve well in water and last longer within food and inside the mouth.

One of the most common odorous chemicals, terpene, is synthesized from isoprene. Isoprene is a very simple chemical with one hydrogen molecule bound to each of the five carbons. Since conjugated carbons are connected in a zigzag manner, similar to the layout of bricks, a surprising number and variety of derivatives can be created, which are the terpenes. Terpenes are included in the leaves and barks of needle-leaf trees, citrus fruits, and flowers. They provide countless important smells of herbs and spices. Because of the high volatility of terpenes, they are the first to make contact with our noses and provide the initial refreshing impression, before the other volatile compounds. These refreshing scents quickly disappear with a little heating because of their high volatility.

Why do plants produce aroma compounds?

Attracting bacteria, insects, and animals

The roots of leguminous plants secrete flavonoid, thus the activated genes then control the growth and functions of root tubercles. Nitrogen fixing bacteria maintain a mutually beneficial relationship with leguminous plants by living inside the root tubercles. Over 80% of plants procure their nitrogen source (i.e., NO_3), which is the most crucial nutrient for plant growth, through this symbiotic relationship.

Plants also attract pollinating insects with their volatile odorants. In order for flowering plants to attract the best pollinating insects at night, volatile substances are more effective than the color or shape of their flowers. This symbiotic relationship with insects plays a crucial role in allowing the flowering plants to overcome the effects of the needle-leaf trees. Flowers use a pollinating system to maintain the quality of their genes. However, fruits lure animals into spread their seeds. Flowering plants dominate this particular world but insects have taken on the role of spreading the pollen by moving around from one flower to another. Before this symbiotic system had developed, plants depended on the wind to spread their pollen. However, insect pollination subsequently became their method of reproduction because of the higher success rate than with wind pollination. After the establishment of the symbiotic relationship between plants and insects, the flowers of plants became much more colorful in order to attract more insects and their pollen became moister in order to stick to the insects' legs better. Thus, colorful *angiospermae*, which interacted with the insects, grew and bred quickly. Flowering plants with colorful flowers multiplied with tremendous speed, while the number of gymnosperms declined.

Fruits have a very unique feature. The tastes of vegetables and meats are softer and more delicate when they are less matured. However, unripe fruits are not tasty. The growth of fruits is very similar to that of vegetables, but they grow at a tremendous rate in the stage when the energy repository cells are expanding. During their most active growth period, melons grow more than $80 \, cm^3$ in a day. The repository cells of fully-grown fruits are the biggest cells in the plant kingdom. A watermelon cell is about 1 mm in diameter. In the growth stage of most fruits, carbohydrates are stored in the cell vacuole as starch granules. Alkaloids and tannins, that is, toxic chemicals in charge of preventing infection and spoilage, and astringent chemicals, respectively, are also stored within the cells. Moreover, various enzyme systems are activated. When all the preparations are complete, the fruit starts to ripen in order to attract animals to spread its seeds. The process of ripening is essentially the progression of self-decomposition. The decomposition of carbohydrates increases the sugar level and softens the texture. The protective chemicals, such as acids and tannin, start to disappear and a uniquely attractive aroma is also produced. The color changes from a typical green color into an easily detectable yellow or red color, to visually advertise itself. As fruits ripen, they actively prepare for their splendid ending as a banquet for animals and humans.

Enzymes oversee this ripening process of fruits and vegetables. The substance that initiates this enzymatic reaction is ethylene (C_2H_4). This hydrocarbon gas (i.e., one of the simplest hydrocarbons) is the trigger that sets off the ripening stage of fully grown unripe fruits. Right before the beginning of the ripening stage, the production of ethylene turns on the switch towards death. When the ethylene triggers this first stage of ripening, it also increases the rate of acceleration of ethylene production and stimulates the respiration rate of the fruit to be 2–5 times faster than before. The taste, texture, and color of the fruit drastically change and subsequently rapidly decline. Fruits with this physiology are called climacteric fruits, which are the part of the reproduction strategy of flowering plants. Tobacco plants produce benzyl acetone to attract bees but also produce the bitter chemical nicotine to chase away bees, resulting in air-traffic control in order for more bee groups to circulate. In some cases, plants use the strategy that the enemy of my enemy is a friend, for example, emitting volatile substances to attract the natural predators of insects that gnaw off their leaves.

As a defensive mechanism

When trees become stressed, they release salicylic acid, an aspirin precursor, in the form of a vapor, into the ecological network. This substance is a signal from the innate defensive system that triggers a series of processes to produce chemicals that animals do not like and cannot digest. It will also be recognized and understood by the neighboring plants so that they too can implement their defensive mechanism. Plants are not mobile, so their only defensive system against insects and herbivores is the production of various chemical substances. Hence high molecular weight polymers are utilized for the plant structures and low molecular weight chemicals are used in their defensive substances.

The degree of spiciness (i.e., spicy heat or pungency) of chili peppers is measured by the amount of capsaicin. The spiciness index is expressed by the Scoville scale, where a rating of 0 is equivalent to a diluted concentration of capsaicin with no spicy taste. Pure capsaicin has a rating of 15 million on the Scoville scale. Habaneros, the hottest chili, has a rating of 200 000 or more. This rate means that habanero extract must be diluted over 200 000 times to reach the level where the capsaicin spiciness is undetectable. Capsaicin is actually a chemical weapon of chili peppers in order to protect themselves against animals. Humans are the only animals that crave the spicy taste of chilis as no other animals like the taste. However, birds have a different taste receptor for the capsaicin. Since the capsaicin does not bind to these receptors, birds cannot perceive the spicy heat. It is a reproduction strategy of chili peppers that they use birds to multiply and spread their seeds to greater distances.

There are also more advanced defense mechanisms. It has been revealed that mustard plants use wasps to kill caterpillars. In 2012, a research team in The Netherlands discovered that the defense mechanisms of plants are much more strategic than we previously thought (Fatouros *et al.*, 2012). The research team of Nina Fatouros at Wageningen University in The Netherlands studied Brussels sprouts. Large cabbage white butterflies (*Pieris brassicae*) and small cabbage moths (*Mamestra brassicae*) lay eggs on the Brussels sprouts, and two parasitic wasps (*Trichogramma brassicae* and *Cotesia glomerata*) parasitize the moth eggs. Brussels sprouts have two methods of defense against the gnawing attacks of butterfly larvae. Firstly, when the herbivorous insects lay eggs on the leaves, the Brussels sprouts modify the wax layer of their leaves so that the eggs cannot hatch properly (Blenn *et al.*, 2012). The eggs of cabbage white butterflies actually react to the particular chemicals that are secreted from the leaves and a result, the leaves dry out within about a day. Thus, the eggs cannot properly hatch. However, when small cabbage moths lay a pile of eggs, the leaves do not dry out. In order to prevent these eggs from hatching, Brussels sprouts use their second method. The sprouts attract parasitic wasps to lay parasitic eggs within the eggs of the small cabbage moths (Fotouros *et al.*, 2012). When cabbage moths lay eggs on the leaves, the Brussels sprouts excrete chemicals that alert parasitic wasps, signaling that "food is present." The female parasitic wasps then gather to lay eggs in the cabbage moth eggs. From the perspective of the Brussels sprouts, they are hiring killers in order to proactively prevent the attack from the herbivorous larvae before they are born.

Attacking tools

Plants also have defense systems to compete with neighboring plants as well as a system for attacking animals. Plants release toxic chemicals from their leaves and stems in order to survive better than the competition, this is called allelopathy. Some herbs and geraniums that are commonly grown indoors for their pleasant aroma have an allelopathic effect. They do not normally give off any odor, but when they are touched by anything, they suddenly expel distinct odors. The toxicity of solanine in potato sprouts and allicin in garlics are all a means of protection. In fact, they even produce antibiotics. Plants utilize a whole range possible methods for the sake of their survival.

The phytoncide that is often mentioned as a beneficial chemical in a forest is derived from the words "phyton" (or "plant") and "cide" (or "to kill"). Thus phytoncide means "sterilizing chemical secreted by plants." This term was first coined in 1928 by Dr Boris P. Tokin, a Russian biochemist. In the west, a Russia-born American bacteriologist, Selman Abraham Waksman, actively studied the therapeutic effect of the phytoncide. Waksman received the Nobel Prize in Physiology or Medicine in 1952 for eradicating tuberculosis with his discovery of streptomycin. This generic name for the antimicrobial substances from plants, phytoncides, includes terpenes, phenolic compounds, alkaloids, glycosides, and others. However, all healthy higher plants have antimicrobial activity and contain phytoncides in some form. The strong antimicrobial agents released by unhealthy or damaged tissues after invasion of pathogens are classified as phytoalexins. Phytoncide is not as strong as antibiotics prescribed to kill pathogenic bacteria, but it is a type of growth inhibiting substance against bacteria to prevent infections.

Coincidental byproducts

All organisms in the plant kingdom produce terpenes. Two isoprenes (C_5H_8) combine to create a terpene (monoterpene, $C_{10}H_{16}$), three combine to form a sesquiterpene ($C_{15}H_{24}$), and four to form a diterpene ($C_{20}H_{32}$). Two sesquiterpene units combine to form a triterpene (i.e., six isoprenes, $C_{30}H_{48}$). Squalene (a triterpene) turns into lycopin with a small alteration. Two diterpenes form a tetraterpene. Carotenoids and lycopene are very important tetraterpenes. Carotenoids participate in the process of photosynthesis along with chlorophyll, and the production of vitamin A. When carotenoids divide into two, they produce diterpenes (e.g., retinol is a carotenoid-based diterpene). A large number of molecules are synthesized from terpenes. They are also decomposed and secreted from plant bodies. The secreted chemical byproducts are the origins of the smells of plants.

Animals generally smell odorants, not produce them

Animals do not normally produce aromas like plants do. Therefore, only limited types of odorous chemicals of animal origins are utilized commercially. The animal-origin raw materials used are musk, civet, castoreum, and ambergris, but only in very expensive perfume products. Since the raw materials are hard to obtain in large quantities, most of them are chemically synthesized for commercial uses.

Animal-origin raw materials

Musk: The external secretion glands of musk deer are located near the male reproductive organ. During the mating season, the musk gland swells to the size of an egg, and secretes a honey-like exudate with a strong smell. The male musk deer secrete the substance to mark their territory and to seduce females from far away. Musk deer are found around alpine valley scrub habitats or forests in the mountains of southern Asia, notably in the Himalayas.

Civet: The civet is a cat-like mammal native to tropical Asia and Africa. Unlike musk deer that have to be killed in order to obtain their musk exudate, killing the civet is not necessary to acquire the musk exudate that is secreted from the perineal glands of both the males and females. African musk civets are reared in farms in Ethiopia for the civet musk and also for Kopi Luwak in the Philippines and Vietnam. Musk civets produce more civet musk when they are under stress. The pure natural civet musk has a very strong smell with a hint of a fecal matter. However, it has a floral odor similar to deer musk when it is diluted. Civet musk has been used as a perfume ingredient for the past 2000 years. It is known that Cleopatra really enjoyed the smell of civet musk.

Castoreum: Castoreum is obtained from the castor sacs of mature beavers. Castor sacs are located under the skin between the pelvis and the base of the tail. Castoreum is extracted with alcohol from the dry matter of Canadian beaver castor sacs. It has a warm and sensual smell of leather when it is diluted, but unprocessed castoreum has an unpleasant odor. It is considered very important in the creation of perfumes to have a leather or chypre perfume scent with a characteristic animal smell. It is also used as a food additive to create vanillin flavor. Until now, chemical synthesis of castoreum has not been successful.

Ambergris: Ambergris is considered as important as musk by some people. It is an important ingredient in perfumes and has also been used as a medicine, a spice, and even as an aphrodisiac to mix with alcohol since ancient times. The possession of ambergris was a symbol of wealth, power, and happiness. Ambergris is a metabolite of sperm whales. It is a biliary secretion of the intestines and has the role of curing wounds in the internal organs after sperm whales have consumed hard foods. The amount of ambergris in a whale can be as much as 100 kg. However, only 0.3% of this is a useful substance for complicated natural scents. It can have an exotic woody and earthly odor, camphor scent, cigarette-like smell, musk, or sea salt smell. Ambergris the size of a soccer ball can be found floating in the ocean and it can very valuable. In 2006, ambergris worth more than a million Australian dollars was found on the seashores of Australia. A family walking down the beach discovered a strange object that turned out to be worth a fortune. Ambergris is normally around $27–87 per gram, but this family found a pile of ambergris weighing 14.75 kg.

Unconditional surrender to pheromones

A pheromone was first isolated in 1956. The German research team of Adolf Butenandt, Nobel Laureate for Chemistry, successfully isolated a pheromone from over 500 000 silkworm moths after struggling for 20 years. The team sampled a specific secreting gland in the abdominal segment of the moths. Just a trace amount of the extract was enough to make male silkworm moths dance crazily using their wings, which suggested that the extract contained a stimulant that induces the excitement of male moths. The research team was able to purify the extract after removing all the unnecessary compounds. They named the stimulating substance bombykol. This was the first pheromone ever identified.

Moths find their mates through pheromones. Out of all species of moths, the silkworm moth has the best sense of smell. In order to seduce the male silkworm moths, female moths release pheromones from their abdomen. The males can detect this smell and then seek their female partner. Silkworm moths that have only a 2–3 mm wing span can detect and locate the pheromone from over 10 km away. Even if there were only one pheromone molecule out of 1 billion air molecules, the silkworm moths would be able to detect its presence. Bombykol molecules from female silkworm moths diffuse into the air and are diluted to a much lower concentration. However, the 4000 olfactory receptors in a male silkworm moth can react with these bombykol molecules, which is how male silkworm moths can successfully find their females even in the dark from over 4 km away.

Is body odor a coincidental byproduct?

Has the olfactory sense been completely set aside from human communication tools? Many scientists point out that such an opinion is incorrect. The human olfactory system is very well developed as humans have more olfactory glands than any other primates. Most body odors are coincidental byproducts of metabolisms. However, some body odors have been created through numerous processes of natural selection in order to deliver special information to others. Such body odors act as a type of signal. They are created from two types of special glands in the skin: sebaceous glands and apocrine glands. Sebaceous glands are widely distributed in the scalp, face, neck, back, chest, and other areas of the skin. Since they secrete oily matter, these glands can cause acne. Apocrine glands are widely distributed around the armpits and genitals. Thick sweat containing musk-scented steroids is secreted from apocrine glands. Body odor is mainly characterized by the smell from the apocrine glands.

There are no attractive body odors that come from the human body. On the other hand, one of the worst body odors is the smell of bacterial metabolites in the wastes from our bodies. The sebaceous glands start to release a type of odorless oil from a young adolescent age. After the bacteria utilize this oil, they produce fatty acids that possess an unpleasant smell. These same bacteria also affect the quality of butter, which is why the smell of sweat is sometimes similar to that of spoiled butter. The unpleasant smell from armpits is mainly a chemical called 3-methyl-2-hexanoic acid (MHA). It was first discovered by George Preti at the Monell Chemical Senses Center in Philadelphia in 1991.

Historically, people have used perfumes to mask their body odors. However, since most people now take showers daily, perfumes are no longer used for that purpose. Some people hold the belief that there is a strong relationship between the use of perfumes and sexual activity and treat perfumes as a kind of weak aphrodisiac. Under such assumptions as these, the premise of the question of "why do people use perfumes in order to cover up their body odor?" might be incorrect. People no longer use perfumes for the purpose of hiding their body odor, but for strengthening their intrinsic body smell characteristics. In reality, the many popular perfumes that are available on the market contain various chemicals that smell similar to body odors.

Most flavors that we enjoy are created by cooking

Most of the natural smells in the world are produced by plants, but the flavors of the foods that we enjoy are created by fermentation or cooking. The only raw foods that have pleasant tastes and flavors are vegetables and fruits. However, they are actually different cultivars that have undergone many improvements to their quality through various breeding technologies compared with their non-domesticated types. Even so, they still have the most natural flavors among the foods we normally consume. The rest of the flavors and odors we are familiar with are obtained by cooking. Richard Wrangham of Harvard University stated in his book *Catching Fire: How Cooking made us Human* that the most important and outstanding invention by human beings is not any tool, language, or civilization, but cooking (Wrangham, 2009). Cooking allowed us to reduce the effort of digestion and chewing by converting raw edible materials into high quality foods that are easily digested. The energy and time that this has saved has been allocated toward the development of the human brain. Cooking has also been a decisive factor in the cultural development of civilization, such as the gender roles.

Is cooking really that important? There is not much difference in the calorie contents of foods before and after cooking. However, the absorption rate (i.e., bioavailability) increases 40–50%. This 40% increase might not sound like a big deal, but the cost and the compensation for the increased digestive function and absorption rate is huge. The surplus of nutrients and its consequence, that is a reduction in the size of our digestive organs and a system with a higher efficiency, are the essential factors of evolution. The cooking process also enhances tastes and flavors as well as the digestive efficiency. It cannot be established whether such flavors from cooking were immediately enjoyed the first time human beings started cooking, but it is very likely that most people gradually began to the cooked flavors because the human body has the ability to remember beneficial foods by their delicious tastes and good flavors.

Nowadays, we cook foods to develop their flavor rather than for digestibility or nutritional value. Cooking clearly creates a variety of chemicals. Aroma compounds are very limited in raw foods such as coffee beans, tea leaves, vanilla, and cacao fruits. However, over 1000 different volatile chemicals have been discovered from roasted coffee beans, 470 from black tea, and 400 from baked bread. Chocolate (cacao) does not have the same familiar taste before processing as the cacao beans have a very strong bitter taste with no other taste or blended flavor. However, to transform such cacao beans into a delicious dessert, they are placed in a basket, covered with banana leaves, fermented, dried, and roasted.

Plants produce odorous materials using enzymatic reactions; humans produce flavors using technology, that is, heat, time, and microorganisms (or enzymes). Plants produce aroma for survival; humans create flavors to enhance palatability and preference. Applying a combination of the processes that will be described next creates popular flavors.

Flavor production by enzymatic or microbial fermentations

Fermented thornback ray (a type of skate fish) are used to make steamed rays, which is a very exotic food in the southern peninsula of Korea. However, it smells of ammonia. Anyone would grimace after tasting steamed rays for the first time, although there is also no doubt that it has an appealing taste because many gourmets like the ammonia smell and search for restaurants that serve well prepared steamed rays. In reality, the ammonia does not just smell strong; it is poisonous. If a strong dose of ammonia is inhaled for a long time, it can cause a life-threatening situation. However, when it is dissolved in alcohol, the ammonia gas is very useful for respiratory stimulation. So why does the thornback ray, unlike other fish, produce ammonia during its fermentation?

Salt-water fish must prevent water loss from their bodies through osmosis into the ocean environment, which has a high salinity. This is why various types of solutes have to be dissolved in the body fluids to increase their chemical concentration. Salt-water fish have been adapted in such a way. However, thornback rays (and also sharks) have evolved a different way of controlling the osmotic pressure by using a lot of urea and urea precursors. When thornback rays have fermented sufficiently, after 2–3 days at room temperature or 1–2 days in the ashes of hay and straw, a considerable amount of ammonia is produced from the decomposed urea. When the fermented thornback rays are steamed, the residual ammonia and unfermented urea stimulate our noses. Fortunately, the concentration of ammonia that can stimulate our noses is very low despite its strong smell, and, therefore, there is no need to worry about its toxicity. An isolated island, Heuksando, the home of thornback rays, is located far away, around 90 km, from the nearest port in the mainland of Korea. Fresh rays are very delicious. However, in the past, the rays had to be fermented in order for people in-land to taste them. Nowadays, refrigeration allows all citizens in the country to enjoy the fish without fermentation, although many of them still enjoy eating the fermented rays with the strong ammonia smell. All fermented food products, not just steamed thornback rays, have the same characteristics. The taste and flavor become richer as the food is decomposed into small molecules through the fermentation process. Globally, most traditional ethnic foods are fermented foods, and all fermented foods are very addictive.

During fermentation, organic acids (lactic acid, acetic acid, citric acid, gluconic acid, and others) are produced in an anaerobic state through the formation of intermediate byproducts of sugars from carbohydrates and various enzymes from bacteria or yeasts. Carbon dioxide and alcohol are produced from carbohydrates. Proteins are hydrolyzed to compounds with a taste such as glutamic acids. Typical fermented products in various cultures include yogurt, cheese, kimchi, soy sauce, soy paste, vinegar, fish sauce, and various alcoholic products.

Fermentation of alcohol, vinegar, and bread: Alcohol is the most symbolic fermented product. Out of all forms of alcohol, wine from grapes is treated as the healthiest product. However, is wine enjoyed for the health promoting effects known as the "French Paradox"? Ingestion of alcohol and the various flavors of wine are the main reason why we enjoy the wine, not because of the health promoting effects, and the fermentation creates a more diverse set of flavors. There are various types of alcoholic drinks depending on the region and the raw materials, and there are many different types of raw materials, molds, and enzymes native to each region other than the variation of the yeasts. In Europe, the byproducts of alcohol and beer production are used for leavening breads.

Although traditionally the yeast harvested from alcohol production was used for bread leavening, yeast exclusively for bread leavening was commercialized in 1868. Yeast produces carbon dioxide to raise the dough during a proofing process. However, it is also a very important raw material for producing esters, alcohol, and organic acids that provide the flavor and texture of the bread. The flavor is also generated by the browning reaction inside a hot oven. This flavor is important, but the total flavor of the bread is the result of a balance of aromas created during the proofing and baking processes. Amino acids extracted from the yeast also provide a characteristic flavor to the bread and aldehydes are also produced by the yeast fermentation. Volatile flavors are created from the thermal degradation of sugars in the dough or from the amino-carbonyl reaction. All of these odorous substances, including alcohol, organic acids, amino acids, aldehydes, and other volatile compounds, are combined to form the characteristic flavor of the bread.

Vinegar is a food additive with a very long history, almost as long as that of wine it is produced from ethanol by acetic acid bacteria. When the wine fermentation progresses further, under air, it turns into vinegar. Europeans considers wine to be the traditional fermented drink but also consider the history of vinegar to be the same as that of wine. After realizing that the sour taste that came from poorly made wine barrels was useful in the preservation of food and had medicinal effects, vinegar was produced intentionally and used widely. Asian countries also used filtered rice wine to produce rice wine vinegar through the acetic acid bacteria fermentation. Likewise, many raw materials can be used for vinegar production. The alcohol and acetic acid fermentations of fruit vinegar, lemon vinegar, corn vinegar, malt vinegar, and so on, are all totally natural biochemical phenomena that even occur in our bodies.

Lactic acid bacteria fermentation of dairy products: The most popular examples of fermented dairy products are cheese and yogurt. The mythical story about cheese production is a common legend among the Arabian traders who traveled through the desert with milk stored inside a sack made of sheep stomach, who discovered that the milk changed into a curd. The development of the flavor in cheese begins with the formation of lactic acid, amino acid, and fatty acids from the sugar, proteins, and lipids in milk, respectively, with the help of lactic acid bacteria and other microorganisms. Pyruvic acid is an intermediate metabolite of sugar, and from this, various flavoring compounds such as diacetyl, acetoin, lactic acids, or propionic acid are created. Furthermore, amines, fatty acid conjugates, sulfur compounds, aldehydes, and ketone groups are created from the decarboxylation, deamination, or degradation of amino acids. Fatty acids are the major flavor substances in cheese derived from milk fats, including butanoic acids, octanoic acid, and decanoic acid. From these flavoring substances, originating from three ingredients (i.e., sugar, proteins, and milk fat), special flavoring compounds such as methyl ketones, secondary alcohols, and lactones are produced. When butterfat is slightly decomposed, it smells like milk, when it is significantly decomposed, it smells like butter, but when it has decomposed completely, it smells like cheese.

Cheese is a fermented product made after the removal of a large amount of moisture, lactose, and whey proteins through the agglomeration of casein proteins, but yogurt is directly fermented milk. Yogurt was not originally a familiar food product in Western Europe and the United States. Until about the beginning of the twentieth century, yogurt was an exotic product of Eastern Europe, Northern Africa, and Central Asia. Western Europeans and Americans were merely curious about it and did not consume it regularly. It became generally accepted after Mechnikov introduced the fermented milk as a secret food for longevity. Yogurt flavors are composed of acids and volatile substances produced from the fermentation of milk, with aldehydes being one of the unique flavors.

In the Far Eastern regions furthest from Europe, there is also a very popular fermented food: kimchi, from Korea. Well-ripened kimchi has lots of lactic acid bacteria, 10–100 times more than other fermented milk products. Moreover, organic acids such as lactic acids and citric acids produced from the lactic acid bacteria provide the capacity for long storage times and also the flavor. Acids such as these are mainly perceived to be sour, but their acidities and sourness actually vary a lot depending on the type of acid, for example, succinic acid can provide umami. Furthermore, the combination of various acids can create a new taste. The flavoring substances from kimchi are also produced by the fermentation process, through molecular changes.

Flavor production by heat processes

The production of coffee from coffee beans and chocolate from cocoa beans involves three major processes: fermentation, drying, and roasting. Each process has a big influence on the flavor of the finished product. The chemical changes during these processes are very complicated and extremely sensitive to the temperature and time used (Figure 5.2).

The flavor of meat is also created by high-temperature cooking. All raw meat has a weak bloody smell. The roasted meat flavor is only

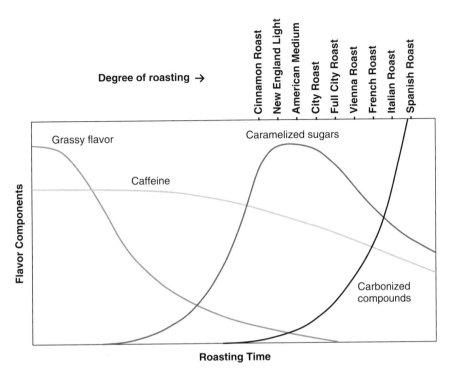

Figure 5.2   Heating decreases the grassy flavor of coffee beans while enhancing the color and other flavors. Modified from (www.therappahannockcoffee.com/roasting/) (accessed 17 July 2014). (See color figure in color plate section).

created when it is cooked. The flavoring substances are low molecular weight chemicals. Roasted meat flavor is created by the decomposition of amino acids and sugars, and their complicated chemical reactions under the influence of heat. The most popular meat flavor and the one in the highest demand is beef flavor. Therefore, many research projects have been conducted to create a beef flavor without actually using beef. For example, some procedures for producing beef flavor were developed after the inclusion of amino acids such as cysteine, protein hydrolysates, or reducing sugars in tallow. Other types of meat flavors can also be created in the same way. To produce a pork flavor, lard has to be used instead of tallow, and to create chicken flavor, chicken fat has to be used. Because the roasted beef flavor is actually a chemical compound produced by various reactions initiated by the heating process, it is possible to make the beef flavors through controlled heat treatment (Figure 5.3). The thermal process produces aromatic substances through the Maillard reaction and lipid oxidation. The combination of chemical reaction products from the Maillard reaction

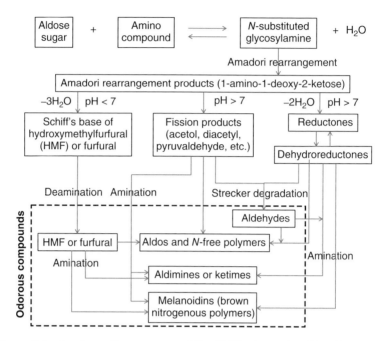

Figure 5.3 A schematic summary of the Maillard reaction. The flavor after cooking is created by the heating of amino acids and sugar.

and lipid oxidation creates a beef flavor. The Maillard reaction starts with sugar, amino acids, and polypeptides present in beef, and is the cause of browning in foods, especially in meat or other foods containing sugar and amino acids, when they are heated above 130 °C.

Under controlled reaction conditions with high heat, the browning reaction flavors are made from at least two precursors. The precursors do not contain any odors, but the delicious smells are produced when they undergo a controlled thermal process such as cooking. Traditional compound flavors are extracted from natural raw materials or from simple mixtures of the natural extracts. The heating process does not play any role in the production of the compound flavors, but heat is involved to initiate the complicated chemical reactions for the browning reaction flavors.

Flavor production by pyrolysis: smoke flavor

The smoking process has been gradually developed over thousands of years. Drying meats or fish in an area full of smoke can help preserve the food for a longer time and also increase the taste of the food. At first this was only a method for preserving food, but it later became a method of providing a flavor. Initially, the smoking process was performed in a kiln without any method of controlling the smoke. Nowadays, smoking chambers burn wood chips with controlled air ventilation and temperature. However, it is still difficult to control the quality of the flavor of the product, because gradually burning the wood chips requires a certain amount of skill to control the temperature of the smoker.

The use of smoke chambers with burning wood chips is relatively inefficient. The smoke chambers accumulate tar-like substances, and particles such as ash on the surface of smoked products can be visually unappealing. Owing to the various problems and technical difficulties of the smoking process, liquid smoking methods are widely used commercially.

Compound flavor: creation of new flavors by mixing various odors

Fermentation and heating both create very rich and excellent flavors. However, we cannot create all desired flavors precisely. Even if we wish to create a strawberry flavor, we cannot make the strawberry flavor through a fermentation or a heating process. It is impossible to imitate a natural flavor by combining any of the other natural substances that already exist in the world. It may be possible to create a cola flavor by combining lemon, lime, and orange oil with spices such as cinnamon, ginger, nutmeg, clove, and coriander. It may also be possible to create a soda pop flavor by mixing lemon and lime juices with vanilla flavor and banana flavor. However, no matter what sort of natural flavors are combined, the flavors of vegetables, fruits, and flowers cannot be created simply through the mixing of natural flavors. The only way to create such flavors is by analyzing the composition of the flavors and compounding each odor chemical one by one. All flavors are a mixture of aromatic chemicals. No matter what the flavor is, as long as we accurately know the compositions of the flavor and all the odorous chemicals can be procured, it is possible to create the flavor by compounding all the odorous fractions together to formulate the exact same compositions. We commonly classify flavors as natural flavors and synthetic flavors, but compound flavors are a more appropriate expression of synthetic flavors. The compound flavors may have synthetic chemicals as well as natural chemicals. In most cases, the compound flavors are made by adding odorous chemicals to a natural extract as a base. A natural flavor retains its complex natural formulation. The natural extract is an extracted material so it is a concentrated material. On the other hand, a compound flavor is created by identifying each fraction of the original natural flavor and mixing all of the identified individual chemicals of the fraction to match the original compositions.

Thus, the final flavoring product is developed through the fractional analysis of the natural flavoring extract, the procurement of all of the single odorous fractions, the development of a base material by combining all of the odorous fractions, and the formation of the products to be of practical end use. Such compound flavors were historically developed for cosmetic products such as perfumes before they were used for food products. Since cosmetic products are not edible, there is a wider range of raw materials available to select from compared with the limited options of edible raw materials available for the selection for food flavors. There are a lot of natural raw materials, but they are not allowed to be used for food flavors or food products before their safety is officially verified.

There are 10 000 food product flavors, but in the end all the flavors in food products can be recreated with less than 1000 odorous chemicals. However, we actually use 2000 types of odorous chemicals and we still cannot recreate all the food flavors because we lack the skill. There are too many misunderstandings and too much mistrust associated with compound flavors. Many people typically say that the synthetic flavoring chemicals such as compounded banana flavors are fake because they are not made from real bananas. So next we will find out how compound flavors are really made and clear up several misconceptions related to them.

References

Blenn, B., Bandoly, M., Küffner, A., Otte, T., Geiselhardt, S., Fatouros, N.E., Hilker, M. (2012) Insect egg deposition induces indirect defense and epicuticular wax changes in Arabidopsis thaliana. *J. Chem. Ecol.* **38**(7): 882–892.

Fatouros, N.E., Lucas-Barbosa, D., Weldegergis, B.T., Pashalidou, F.G., van Loon, J.J.A., Dicke, M., Harvey, J.A., Gols, R., Huigens, M.E. (2012) Plant volatiles induced by herbivore egg deposition affect insects of different trophic levels. *PLoS One.* **7**(8): e43607.

Wrangham, R. (2009) *Catching Fire: How Cooking Made Us Human.* Basic Books, New York.

6 Technological Advancements Brought about by the Love of Flavors

La découverte d'un mets nouveau fait plus pour le bonheur du genre humain que la découverte d'une étoile. (The discovery of a new dish confers more happiness on humanity, than the discovery of a new star.)

Jean Anthelme Brillat-Savarin (Physiologie du goût, 1825)

People have appreciated many different kinds of flavors for a long time. The love for spices during the Middle Ages is probably one of the best examples. Black pepper is now a very common spice on dinner tables all over the world. However, spices including peppers, cinnamon, cloves, nutmeg, and ginger, were luxurious items only available to a few rich people during the Middle Ages. Dried black pepper was so valuable during medieval times that 1 pound could free a servant who belonged to a lord of a manor. The desire for spices gradually increased and the enormous demand after the fifteenth century led to the Age of Discoveries.

The legendary figures that spread the mysteries of the Orient, such as Marco Polo and the mythical Sinbad, are all associated with spices. In addition, the Bible also mentions the Queen of Sheba, who visited King Solomon in the tenth century B.C. She presented him with gold, precious stones, and most importantly spices. These spices were substances added to foods or drinks to affect their taste and flavor. As we can see from this story, gold and spices have been closely

How Flavor Works: The Science of Taste and Aroma, First Edition.
Nak-Eon Choi and Jung H. Han.
© 2015 John Wiley & Sons, Ltd. Published 2015 by John Wiley & Sons, Ltd.

entwined since ancient times, as they have remained synonymous with invaluable riches throughout history. Even during the Roman Era, which was known for its abundances and wealth, only a select few people had access to spices. Various spices appeared on the tables of the aristocracy, the wealthy, and the clergy. Guests thus judged the richness of their hosts based on the spices served, making spices the true symbols of taste and wealth. Aristocrats at the time preferred the most expensive condiments because the cost of the meals influenced the value of the food and the class of the dinner table while also adding flavor. These aristocrats explored new lands in order to obtain more tasty spices. In the fifteenth century, Venetian merchants monopolized the spice trade and left no room for other countries to share in the immense profits that were achievable. Other countries began to seriously consider the possibility of pioneering a new route to India, particularly a sea route circumnavigating Africa. Prince Henrique of Portugal, Duke of Viseu, organized a fleet of sturdy merchant ships, and thus began the Age of Discoveries. Pope Urban II, who initiated the Crusades, was a stingy pope, but he was well known for enjoying meats. With the spread of Islamic power in the east of the Mediterranean, the importation of peppers was critically affected during his papacy. Peppers became scarce and their prices rose astronomically. Although Pope Urban II started the Crusades to recapture the Holy Land, after pilgrimage to Jerusalem became difficult because of the expansion of Islamic influences, he may not have considered that others saw it as a solution to the shortage of peppers. Thus, Pope Urban II probably did not know that the goal of the Crusades might have had the hidden agenda of securing the importation of peppers.

Nutmeg and cloves were more valuable than peppers. They originated from the Moluccas (Maluku Islands, which today is Indonesia) and had to go through a distribution channel of 12 steps to reach the consumers in Western Europe, with the price doubling after each step. Magellan traveled the world in search of nutmeg and cloves. Spices were beginning to awaken the senses of the people in Medieval Europe, where religious ideologies dominated and the body and emotions were looked down upon. However, the desire of the senses to consume spices was slowly dawning in Europe. Although Zheng He of China explored the world from 1405 to 1433 with a larger fleet (over 27 000 crews in 67 vessels) earlier than any European fleet, his travels were cut short due to the failure to find items worthy of trading.

Even though spices have played such significant roles, they do not possess any special functions. Spices simply have chemicals that possess more odors compared with the thousands of other plants on Earth: they just have different odorous chemicals. This statement does not mean that the spices have a special component. Spices are simply composed of various odorous chemicals in different amounts. Sometimes one odorous chemical may provide the main odor, such as in the case of cinnamon, cloves, or anise. However, most spices contain several odorous chemicals. Fruit flavors are the same as spice flavors. Every flavor is nothing more or less than a combination of odorous volatile chemicals. Their characteristics are simply due to the differences in the types and amounts of the odorous chemicals.

Naturally scented plants such as herbs and spices go through minimal processing, such as drying and grinding, prior to use in cooking. Sometimes, they undergo additional processes in order to be used as an ingredient for compound flavor. Previously, the aromatic compounds extracted from natural herbs and spices performed the major role in compound flavor. However, since various synthetic odorous chemicals have been developed, the usefulness of the flavor combination with natural extracts and their fractions has decreased, and, consequently, their productions have decreased as well. The raw ingredients for compound flavors are numerous. They are produced from different parts of herbs or spices. Naturally grown plants show some variation in their flavors according to their breed and cultivation locations. Their flavors are also different depending on the parts of the plants from which they are extracted, such as the fruit, flower, branch, or leaf. Even from the fruits, the flavor can vary depending on which area of the fruit it was extracted from, such as the peel, flesh, or juice. Extraction methods add further variations. These include compression, steam distillation, and solvent extraction, all of which lead to different results for the flavors. In conclusion, the flavors and the purpose of the extracts vary depending on the final product types, including essential oils, essences, or oleoresin.

Approximately 70 species of herbs and spices have been identified throughout the world, but the majority of them are only used in local cuisines. There are just a few herbs and spices that are commonly used around the world. Black pepper seed is the most important trading item, with respect to price and quantity, other than grains, legumes, or oil seeds, followed by sesame, mustard, and other herbs. The most frequently traded dried spices, excluding herbs, are black pepper, cloves, nutmeg, mace, chili, cardamom, cinnamon, cassia, ginger, and allspice (Jamaican pepper). These ten spices constitute 90% of the global market.

Why do people combine flavors?

Spices are used for preparing meals, while flavors are used for making beverages and ice cream. Even though the methods of how to use them and the final products are different, spices and flavors are essentially used for the same purpose, which is to enhance and diversify aromas to increase the preference of the food and beverage products.

Flavoring
  ○ Gives characteristics by adding flavors to a tasteless product.

Enhancement
  ○ Emphasizes the original aroma in the food.
  ○ Supplements the aroma of the food lost during the processing.
  ○ Decreases the variation in quality due to seasonal or environmental factors.

Modification
  ○ Changes the undesirable original aroma to increase the preference.
  ○ Reduces unpleasant aroma, such as during heating, fermentation, or the smell of fish, to increase preference.

The primary purpose of using flavors is to add or enhance aromas. However, the modification of aromas is difficult. As a result, it is easier to add the appropriate flavors to tasteless or odorless base materials. It is also easier to simply enhance the flavor of a product. However, during product development projects, most new functional foods and nutritional supplements possess bitter tastes or undesirable off-flavors, which result in a reduced preference for the products. Flavors are expected to solve these problems, but there is no technology to mask an off-flavor or bad taste with other flavors. Simply by adding flavors, a new good aroma is incorporated along with the pre-existing bad taste and off-flavor, which compromises the food product quality by diverting the attention of the consumer from the undesirable taste and flavor.

When considering the usage of flavors, compound flavors seem to be frequently utilized. Furthermore, there are many people in the food business who believe the incorporation of flavors may solve all of the problems related to the flavors and tastes of their products. However, their expectations are not practical. According to Fidelity (2013), the US market capital for the beverage business in December 2013 was $824.83 billion; the food product manufacturing business was

$771.00 billion; food services and staple retailing were $704.08 billion; household products were $373.72 billion; personal care products were $168.65 billion; and tobacco was $391.31 billion.[1] According to the USDA (2012), industrial sector values to 2011 GDP are $740 billion for agriculture and related industries, $340 billion for food services and drink business, and $200 billion for the food, beverage and tobacco manufacturing industry.[2] All of these industries use flavors and fragrances. The success of the food business in particular depends on taste and flavor. The food manufacturing and services industry, including tobacco, constitute $540 billion together; therefore, the market size of food flavoring agents might also be considered to be very significant. However, the flavor and fragrance business in the United States is only $5.7 billion, according to the business report of The Freedonia Group (2013), which is just a small portion of the consumer products and goods market. The market size of food grade flavors and fragrances for the food and beverage industry is part of this $5.7 billion. Since the cost of flavors is relatively expensive, $5.7 billion is not particularly high. This shows that the circumstances where flavors can be used are very limited.

In all countries, the beverage industry uses the most food flavoring agents. About 30% of all food flavoring agents is used in beverages. The total sales of the beverage industry in terms of currency, as well as the total production in mass, are huge, since the main component of beverages is water. The amount of flavoring agents is proportional to the weight of the product. The composition of flavors used in chewing gums is very high, but the unit quantity of chewing gums is very low. Therefore total amounts included in the chewing gums cannot compete with the quantity of flavors in beverage products.

Because most flavors are highly volatile, compound flavors cannot be used in bakery products, which utilize high-temperature processing. Perfect sealing for the packaging materials and during the packing processes are also required, because the volatile flavors easily permeate into the atmosphere during storage. Products that require heating processes before consumption cannot incorporate flavors either. The flavors in frozen foods vaporize the moment they are heated. Owing to these restrictive conditions of flavor use, the

---

[1] Market Cap is the sum of the market value of each company assigned to the applicable GICS sector or industry. Market value or capitalization is calculated by multiplying the number of common shares outstanding by the market price per share at the end of each trading day.
[2] USDA Economic Research Service data based on US Department of Commerce, Bureau of Economic Analysis.

number of products containing flavors is very small. Nevertheless, if there are further advances in flavor technology, the number of products using flavors will increase.

How many flavors are there in the world and how many ingredients are required to make all of these flavors?

How many flavors are there in the world? There are many common flavors, such as apples, strawberries, roses, and milk. Simple counting the different flavor categories could be endless. Then, for example, there are also different flavors according to the particular cultivars and to the ripening stages of the apples. It may be impossible to count every type of flavor. Therefore, the types of flavors can be dependent on how many we can count. When asked to categorize types of flavors, one person might say 30 000, while another might just say thousands. In 2004, when Linda Buck and Richard Axel received the Nobel Prize for their discovery of olfactory receptors, the Nobel Foundation published a press release entitled, "People recognize and remember about 10 000 different odors." However, this number did not originate from the research of Buck and Axel. It had been tossed around for years by other scientists from Arthur D. Little, Inc., which has not been proven experimentally or scientifically (Gilbert, 2008). Back in 1927, Ernest C. Crocker and another chemist, Lloyd F. Henderson at Arthur D. Little, were struggling for an objective way to classify odors. They settled on a method in which odors were classified into four elementary odors and each odor was rated by its strength on a scale of from 0 to 8. Given the mathematics of their rating system, it was theoretically possible to discriminate 6561 ($= 9 \times 9 \times 9 \times 9$) different odors. The math is watertight, but the outcome is highly dependent on the initial assumptions. If Crocker and Henderson had used five elementary odors and a 0 to 10 rating scale, the estimate would have been 161 051 ($= 11 \times 11 \times 11 \times 11 \times 11$) different odors. Harvard psychologist Edwin Boring was a fan of the new system, but he believed the rating scale should have fewer steps. He did some calculations and decided that the number of distinguishable smells was somewhere between 2016 and 4410. Discussing this work years later, Ernest Crocker generously rounded up the estimate to 10 000 odors. His colleagues took this number and ran with it. In the end, it appears that no one has ever attempted to count how many smells there are in the world. Estimates of odor diversity either lead to a dead end or to Ernest C. Crocker. The comfortable, often-cited figure of 10 000 smells is, from a scientific perspective, utterly worthless. So this is the story of the 10 000 smells theory (Gilbert, 2008).

How many odorous chemicals are needed to make all of the aromas of plants and foods? Coffee flavor alone has 1000 odorous chemicals. Common food flavors have 400 chemicals. Fruit flavors are relatively simple with over 100 odorous chemicals. Thus, if 100 chemicals exist in each food item, 1 million odorous chemicals may be needed for nearly 10 000 foods. However, this many substances are not required in order to consider all food flavors.

There are roughly 400 000 odorous chemicals in the world. In 1974, 2600 odorous chemicals could be found in foods, but in 1997, 8000 volatile chemicals were identified in foods. Therefore, approximately 10 000 chemicals constitute all the odorants in the world. A smell is like a color. Humans have only three visual receptors, but his does not mean that we can see only three colors. It means that there are three primary colors and a combination of these three colors will result in an infinite number of different colors. Color monitors use three RGB (red, green, blue) colors in 256 gradients to create 160 million colors.

A combination of various odors can result in totally different aromas. If completely distinct odors are mixed, they are not sensed as separate odors, but are perceived as an entirely different but whole odor. As a result, less than 15% of people can identify at least one original odor from a mixture of three odorous chemicals. Perfumers and flavorists cannot identify more than three components from a mixture. Even after training, the specialists are able to identify only four components from a mixture when the test is set at its easiest level. There are three primary colors because there are three color receptors. In the same way, there are 388 primary odors because there are 388 olfactory receptors. Therefore, at least 388 odorous chemicals are needed to explain the full spectrum of odors. People do not use just three paint colors because there are three primary colors but they use 12- or 24-color palettes. There is only one sweet taste receptor, but there are 20 different sweeteners. If ten odorants can bind to one olfactory receptor, nearly 4000 odorous chemicals are reactive to 388 olfactory receptors. If five odorants are assigned to one olfactory receptor, nearly 2000 odorants are needed. All the tastes and flavors in the world can be created theoretically through the combination of a number of odorants, this being less than 10 000 but more than 1000.

How many odorous chemicals are needed to create a tomato flavor?

The first step in making a compound flavor is to analyze the natural source. To make an apple flavor, we must examine the most flavorful apple, and to make a tomato flavor, we must initially examine the best tomato. Since flavors are volatile components, gas chromatography (GC) can reveal the full chemical compositions. Equipment such as a GC-mass spectrometer can establish the identities of these components. This is the easiest way to reveal the secret of a flavor.

Does this mean that flavors can be created through analytical support, using the right equipment, without the help of a perfumer or flavorist? Not at all. All identified volatiles are not flavoring chemicals, and the combination of all identified volatile chemicals to create the apple flavor lead to cost and safety issues. The first thing to consider is the concentration levels of the volatile chemicals, based on their threshold values. The threshold value is the lowest concentration at which a stimulus can be detected and recognized (Table 6.1). The thresholds differ by factors of a million, depending on the odorous chemicals, so they must be considered to be more important than the ratio of volatile chemicals in a flavor.

Acetone, which is used as nail polish remover, has a pretty strong smell. However, the aromatic compounds in foods are stronger than the smell of acetone. Menthol, a compound in common mint candy, is 375 times stronger and butylmercaptan is 50 000 times stronger than acetone. This shows that threshold plays a more important role than the amount of an odorant. The great differences in thresholds of odorants do not mean that some substances possess unique chemical strength or functionality for the human body. The only way to explain the differences in the threshold is that the lower threshold chemicals bind stronger to the G-receptor. Also, the magnitude of an olfactory sense varies depending on the molecular structure of the odorants,

Table 6.1   Threshold Values of Example Odorants; a Lower Threshold Indicates a Stronger Odorant

Odorous Chemical	Threshold (ppb)
Methanol	141 000
Acetone	16 000
Formaldehyde	870
Menthol	40
Butylmercaptan	0.3

not on whether the odorant is natural or artificial. Generally, a natural odorant is perceived to be stronger than its synthetic odorant because the natural odorant does not have optical isomers.

Professor Harry Klee at the University of Florida collected 278 species of tomatoes grown in the United States and asked 170 trained researchers to classify the taste of each tomato (Tieman *et al.*, 2012). The sensory panels were divided into 13 groups and each individual tested 4–6 tomatoes to examine sweetness, sourness, saltiness, bitterness, and umami. The research team analyzed all the volatile chemicals from the tested tomatoes and then studied how each volatile chemical in the tomato was perceived. The results showed that out of 400 identified volatile chemicals, 27 volatiles contributed to the tomato flavor. This confirmed that only a small portion of the volatile chemicals function as characteristic odorants and can be perceived by olfactory systems. Perfumers and flavorists already knew this fact, as their role is to analyze complex molecular structures of natural volatile chemicals, identify the primary odorants, and only use these primary odorants to recreate the same flavor as the original. In some cases, such as bananas, one volatile chemical could be decisive in characterizing the smell, however, in most cases, 10–30 ingredients are sufficient to recreate a flavor. The reason why compound flavor is possible is because a small number of volatile chemicals are sufficient to create a specific flavor.

Compound flavors use volatile chemicals that are no different to those from natural sources as most of them are identical to the naturally existing odorous chemicals. They are not extracted from natural resources, but are chemically modified from naturally existing common chemicals.

The most popular flower throughout the whole of history is probably the rose, the queen of flowers, which has rich and complex fragrances. This cannot be imitated naturally because their scent contains 400 different volatile chemicals. To make 1 L (0.9 kg) of rose oil, about 3500 kg of rose petals are required. A trained worker can harvest 6 kg of rose petals an hour but they can only work early in the morning because roses wilt when it is too hot. So working for 5 hours a trained worker can harvest 30 kg of rose petals, from which just a few drops of rose oil can be extracted. Only 0.2 g of rose oil can be extracted from 1000 rose blossoms. As demonstrated, natural fragrances exist in small amounts. However, such small amounts are strong enough to give off the rich scent of many roses.

People are constantly investigating odorous molecules in order to find the chemicals that fit perfectly with the olfactory receptors, because succeeding at this research will allow them to recreate every odorant with the least number of volatile chemicals. However, the exact relationship between the molecular structure of the volatile chemicals and the receptors has yet to be fully explained. There are similar molecules that give the completely opposite odors because they have different osmophores (a side group in the main common chemical structure), while some molecules with totally different molecular structures give similar odors. Large research funds are required to synthesize one new odorous compound. If we had the technology to allow people to estimate the odor from the molecular structure, all perfume and flavor research and development would be very efficient. However, such a tool does not yet exist. In pharmaceutical science, there have been great advances in the technology to predict the medicinal effects of a chemical from its molecular structure, but in the field of flavor chemistry there is still a long way to go. Nevertheless, the following are some characteristics that are known about the molecular structure of odorous chemicals.

- All odorous molecules are composed of five atoms: carbon, hydrogen, oxygen, nitrogen, and sulfur (C, H, O, N, and S).
- The odorous molecules must be volatile in order to reach the olfactory receptors in the nasal cavity, and lipophilic, with a slight hydrophilicity. The molecular weights are less than 300 (the average is below 200) because a lower molecular weight increases the volatility. There are generally 4–16 carbons in volatile chemicals. The molecules with about 8–10 carbon atoms have the most elegant odor. The fewer carbons there are in the molecule, the more it expresses a strong but short-lasting odor, while more carbons exhibit a more complicated and long-lasting odor.
- Odorous molecules are nonpolar. If the molecules have polarity, there will be intermolecular interactions, which reduces their volatility. If there is an oxygen atom at the end of the molecule, through a hydrogen bond, the volatility will decrease but their receptor binding affinities will increase.
- When the molecular structure is elongated, the time the odor lasts will increase. Typically, the addition of one carbon atom doubles the duration of the odorant.
- A double bond changes the shape of the molecule, and, thus, the position of the double bond influences the spectrum and strength of the odor.
- Hydrocarbons exhibit the strongest odor when there are 8 or 9 carbon atoms in their molecules. More than 16 carbons eliminate the odor of the molecule.

- Ketones (=CO): many odorous molecules contain ketones.
- Alcohols (—OH): about half of all alcohol molecules give off a woody smell. C4 and C5 give off a fusel oil odor, and the odor reaches a peak at C8. Above C14, they become odorless.
- Aldehydes (—CHO): many aliphatic compounds contain aldehydes. C4 and C5 smell butter-like, while C8–C12 smell floral and butter-like. C16 is odorless. Aldehydes of the terpenes give off a citrus flavor. In general, aldehydes give off a stronger odor than alcohols.
- Esters (—CO—O—): an ester is created through the condensation of alcohols and carboxyl acids; most esterification products from short-chain fatty acids (<C6) and short-chain alcohols have an odor and play a critical role in fruit flavors.
- Lactones (—CO—O—): similarly to esters, many lactones have fruit flavors.
- Nitrogen compounds: in many cases, they exhibit animal smells.

Perfumers and flavorists create flavors

Perfumers and flavorists are specialists who mix odorous chemicals to create a preferred aroma. Perfumers make perfumes or household products using only their nose. These products are not restricted by the raw odorous chemicals and as to whether they are natural or synthetic. They only focus on how the olfactory senses are stimulated to make consumers feel good and satisfied. Even if an aroma does not exist in nature, as long as it pleases consumers, the aroma is successful commercially. Therefore, creativity in designing a new aroma is an important skill set for perfumers.

Flavorists develop flavors for food and beverage products, but they must use both their mouth and their nose to examine the ways the chemicals bring out certain flavors with foods. Therefore, flavor ingredients must also be as safe for consumption as the food ingredients. People are very conservative about flavors in their foods and are reluctant to try new ones. Therefore, the general objective for flavorists is to recreate flavors that exist in nature. At one time people were open-minded with respect to new flavors, and scientists invented flavored beverages, such as cola and soft drinks. The unique flavor of cola comes from the combination of cinnamon, ginger, nutmeg, clove, and coriander, with citrus flavors such as lemon, lime, and orange. Colorless soft drinks, such as 7 Up, Sierra Mist, or Sprite, are actually simple combinations of lemon and lime flavors with sweeteners. However, most consumers do not recognize the citrus flavors in the soft drinks. Instead, they label this lemon–lime flavor as the soda flavor, named after the carbonated soft drinks. Cola and soda flavor are the most abstract flavors, because their names do not specify any natural of the flavors that they contain, even though they are very popular flavors in ice cream or frozen dessert products. Soda flavor is a combination of mainly lemon and lime flavors with small amounts of banana and vanilla flavors. However, there are very few people who know that soda flavor is a mixture of fruit flavor with vanilla flavor.

It is not an easy task to imitate natural properties. There are more than 100 volatile chemicals that contribute to strawberry flavor. However, each individual odorous chemical does not smell exactly like strawberries. Instead, they are weak, very strong, or similar to something else, or remind consumers of someone or something through their impressions. Nevertheless, once all of these odorous ingredients are combined, exactly as in the formulation of natural strawberry, then the smell is like strawberries. What people perceive as a strawberry is the result of a harmonious combination of different odorous chemicals that stimulate our sensory organs and our brains. Natural fragrances are already a compound flavor of many naturally existing odorous chemicals, but artificial compounding is a series of trials and errors, collecting and mixing various odorous chemicals to reach an acceptable harmony.

Olfactory training: flavorists must first distinguish odorous chemicals before creating compound flavors

Perfumers and flavorists create numerous fragrant products from hundreds of odorous chemicals. Do they possess keener noses than other people? How do you become a perfume specialist? A person who is suitable to be a perfumer is not a person with a precise sense of smell, but actually someone who feels uninhibited delight and happiness from interacting with and the making of fragrances. The food and perfume industry do not fill their research and development positions with people who have very sensitive olfactory or taste senses. On the tongue of a less sensitive person, there are 100 taste buds per square centimeter, the tongue of an average person has about 200 taste buds, while a sensitive person has about 400 on 1 cm^2. A sensitive person might seem more compatible with the job, but this person is also more sensitive to bitterness, which makes them unsuitable. An average person, on the other hand, enjoys a wider spectrum of foods more deeply than the sensitive person. Thus, average smell and taste sensitivity are sufficient to be a perfumer or a flavorist. In other words, passion and training make a flavor expert rather than exceptional sensitivity. Regular people and experts sense the same flavor to the same extent, but the expert is trained with the cognitive ability to utilize the sensory information better. A trained perfumer classifies new aromas and identifies the unique note of an odor. A trained wine connoisseur recognizes the type of grapes and the vintage year of the wine. Thus all of their abilities depend on trained memory rather than their noses. In other words, the strength of experts comes from intellectual ability, not their sensory ability. Their intellectual abilities can be achieved through routine practice and systematic training.

In the mandatory wine tasting exam to become a master sommelier, the examinee must distinguish and describe six bottles of wine of unknown origin in less than 25 minutes. Four minutes are assigned to taste each bottle and 10 seconds are given to the examinee to explain the type of grapes, area of production, and vintage with accurate substantiation.

"I was concentrating on distinguishing the last bottle of the six samples. Time kept passing by. I looked at the tear of wine (i.e., wine legs) in a glass, savored the aroma, and tasted a sip. At first, it felt like a mixture of an over-ripened Cabernet and a Shiraz. The origin of production seemed like Australia. However, something made me feel unsure. I thought about all the features of the wine again. It was not sweet, it was heavy, high-proof, and had a zippy tang. And it tasted a little bit like raisins. At that time something lit up in my head. 'Oh, this is Italian Amarone!' It was that moment when my investigation and concentration came to fruition." This is an anecdote describing the wine identification process by a participant of the master sommelier test. The answer did not come from the nose or tongue, but from related information.

Sommeliers easily identify the unique characteristics of each wine, but they do not have a more advanced sensitivity than others. Wine experts say that they are no better than normal people at tasting wine components such as tannin or alcohol, as even a beginner can recognize whether one wine is different from another. The wine classification skill of an expert is only valid when the wine characteristics match the information that the expert has acquired. If the characteristics are fabricated, they cannot identify the wine correctly. For example, experts sometimes report a rich red wine flavor from a red colored wine that was made by adding red dye to white wine.

Thus, wine experts do not possess special tongues or noses, but it can be said that they do have special brains. The fMRI brain images of sommeliers show that their brains act very differently from ordinary brains when they taste wine. An Italian radiology research group (Castriota-Scanderbeg *et al.*, 2005) showed activation of the cerebral network in the left insula and adjoining orbito-frontal cortex in sommeliers. Both areas have been implicated in the integration of taste and smell. In addition, other activation was found bilaterally in the dorsolateral prefrontal cortex, which is used for high-level cognitive processes, such as working memory and selection of behavioral strategies. Increase in activity in these parts of the brain makes the sommelier perceive flavor and taste more accurately. The left hemisphere of the brain is responsible for analytical function. Thus, the increased activity of the part of the brain in the left insula and cortex of sommeliers means that the sommeliers experience taste a little more intellectually than normal people. In addition, there is more activity in the part of the brain that is related to advanced cognitive abilities such as memory, language, and decisions when the sommelier swallows wine. Increased activity in this part of the brain further improves the analytical experience of the sommelier. These brain radiology research results adequately explain the true techniques of sommeliers. Wine experts have developed conceptual knowledge about wine as well as having enriched their vocabularies to describe the taste and flavor of wine. There are about 500 rich and expressively descriptive lexicons that describe the taste and flavor of the components of wine as well as the texture inside the mouth. Through constant practice of describing the effects of the components on olfactory and gustatory senses, sommeliers have categorized countless distinctive wines.

Furthermore, extensive knowledge of the types of grapes and production process technologies allows the sommeliers to identify the flavor characteristics that vary considerably depending on these two parameters. For example, most sommeliers are very familiar with the volatiles of wine produced through the malic acid fermentation reaction. In a sampling test, a sommelier is likely to recognize the volatile flavor of a malic acid fermentation reaction, and then move on to the next step of paying attention to the flavor that is associated with this, as the result of lactic acid fermentation, such as that in yogurt and sauerkraut. Through these ways of categorized thinking of the chemical and fermentation reactions, sommeliers easily differentiate and recognize various wines during the sampling test. Unlike beginners, sommeliers judge the similarity between wines based on the already standardized categories of wine. Sommeliers' wide knowledge allows them to easily identify, categorize, and remember the various types of flavors, which are determined by the types of grapes and the processes by which they are produced. Without this knowledge and training, even the wine enthusiast will have a hard time distinguishing various wines.

In conclusion, there is no significant difference between the sensing abilities of wine experts and normal people. Sommeliers do not necessarily have better noses or tongues. A wine expert's technique comes from conceptual knowledge achieved through training, concentrating on the combinations of flavors, and the utilization of descriptive language. A perfumer is also the same as the sommelier. A person with a normal olfactory sense can make a perfume. What makes a perfume expert is the special intellectual ability and cognitive process. Perfumers and flavorists simply have a better imagination than olfactory sense, which allows them to remember the smell of a certain food and imagine how the flavor will be changed after mixing odorous chemicals. The response of their brains to the flavors is enhanced after consistent training of the cognitive ability. Like a sommelier, the brains of perfumers and flavorists have more activity in the orbitofrontal cortex of the frontal lobe area, where cognitive judgment is involved. This brain reaction pattern reflects that the experts perceive aromas more analytically.

How does a perfume master remember odor characteristics and the names of the countless fragrant chemicals? If a normal person saw the hundreds of strange and sometimes unpleasant fragrant chemicals bottled on the shelf in a perfumer's laboratory, they would have no idea what was going on. A perfumer must learn the special cognitive ability of making new intellectual territory and then how to fit a new fragrance into that territory. Thus, to become a perfumer, you must not learn how to smell but how to think.

The first step in the process is mastering the most common odorous chemicals with a training matrix chart. Each column has a group of odorous chemicals, while the rows have odorous chemicals within each group. At first, the trainee must master the differences in the smells between each group, and then learn to compare and contrast within the same groups. For example, to learn the tangerine group, the trainee memorizes the smell of lemon, Bergamot orange, mandarin, orange, blood orange, grapefruit, and lime. The important thing at this stage of training is to create a personal impression of each component. Creating a personal impression plays the most vital role in remembering the tiny differences during the later process of making perfumes. The trainee must identify over 300 odorous chemicals to begin to become a perfumer. An expert perfumer can be familiar with 500–2000 chemicals, all available at the work place, as well as being able to grade each of the odorous chemicals.

In the next step, a trainee who has completed the memorization of odors of all the basic chemicals learns how to think like a perfumer. When an expert is analyzing a new perfume or inventing a new fragrance, the person does not think about individual components. Instead, they think about a typical combination of chemicals, called the harmony. Harmony refers to the combination of particularly well-matched odorous chemicals, and is the basic principle of fragrance production. The perfumer organizes the complex olfactory sensory world into a few fragrance groups to be able to handle them easily, because the objective of a perfumer's work is not to remember the chemicals but to recognize the odor patterns. The mental map of the perfumer is well organized through autonomous details.

Compounding flavors: aromas are completed through imagination

Thierry Wasser, Grand Parfumeur at Guerlain,[3] visited Seoul, Korea to investigate the Korean perfume market and search for new fragrance sources from Korea. A newspaper reporter (Kang, 2009) interviewed him to understand his position within the Guerlain company and answer some questions on the unusual but attractive job of a perfumer.

Is the olfactory ability the most important skill for a perfumer? When you were a student, did you take any screening tests for your smelling ability?

"It's true that the position of the perfumer is about smell; however, the olfactory ability can be improved by training and education. On the contrary, Mr Jean Hadorn, the Director of the perfumery school at Givaudan, thought that my sensitivity and creative thinking would be helpful in creating new perfumes."

---

[3] Thierry, is a contemporary perfumer who, as of 2008, was appointed as the in-house perfumer at Guerlain. Prior to this he worked with the fragrance firms Firmenich and Givaudan. After attaining his Federal Diploma of Botany at age 20, Wasser attended the Givaudan perfumery school and was promoted to perfumer at 24, after a period of time as an apprentice. He joined Firmenich in 1993. There he collaborated extensively with Annick Ménardo on numerous fragrances. His work includes collaborating with Jean Paul Guerlain in the creation of perfumes. (From Wikipedia, the free encyclopedia.)

In order to increase the precision and accuracy of the olfactory abilities, most perfumers pay considerable attention to taking care of their nasal conditions by abstaining from garlic or smoking.

"I smoke and eat garlic. Smoking is an unhealthy habit. However, as long as smoking and eating garlic does not interfere with the condition of my health that is required for my work, I don't think it affects the perfumer's ability to smell. A perfumer's ability to smell can be developed through training."

If an outstanding sensitivity of smell is not important, then what is the most critical skill for a perfumer?

"It's an outstanding brain and stamina. Right now, there are almost 3000 basic odors used in compounding perfumes and fragrances. A perfumer must know all of them. He or she must identify and remember the smell from wherever and whenever, beyond simply knowing them. In that way, he or she can figure out what to mix, how to mix, and in which sequence the fragrances must be mixed in order to match the one designed in their head. I can do that, of course. Anyway, a perfumer must be smart to do this. Therefore, an excellent perfumer is someone who controls the process of such steps as selecting the odors for the production and so on, while specifying the abstract odor profiles designed by their imagination into a real fragrance product. Creativity is the most important competency for a good perfumer."

So through that level of experience and training, the perfumer can create the complicated formula that is the completed ingredient list for a product?

"No, it doesn't mean that. Perfumery is not a science. In fact, I don't even want to use the word 'formula'. A masterpiece painting is not just a good combination of different paint colors. A meal well prepared by a chef is the same. A recipe that has all the steps for preparing a food is not just a list of tasting and flavoring ingredients for mixing. In that sense, I'd rather call the blueprint design of perfumes a recipe than a formula."

I heard that a majority of perfumers are smart people with a chemistry major.

"Once you start work in the perfume industry, you'll realize that the required chemistry knowledge is very basic and it may be sufficient to learn on your own through your work experience. I can't understand the reason, but famous perfumery schools still pick students who majored in chemistry or a related subject at college."

As important elements, health and stamina were a little unexpected for me.

"When Jean-Paul Guerlain, who's like a teacher to me, is out searching for new fragrance ingredients, it's very similar to a scene from the movie Indiana Jones. In work boots, he actively travels the world in search of ingredients for a new smell. To do this, good stamina is very important for an adventurer. For this reason, most perfumers were men in the past. The male to female ratio is about 50:50 right now, but regardless of gender, exceptional body health and stamina are very important."

To become a perfumer, a heavy smoking habit and age do not matter

It could seem that smoking would be a taboo for perfumers, who must maintain a relatively keen sense of smell. Although smoking is known to diminish the sense of smell, there is no scientific evidence to support this. Some research has revealed an adverse effect of a smoking habit on the sensorial sensitivity, but many other research results, including some recent ones, say otherwise (Gilbert, 2008, p. 55). An Australian research experiment with 942 subjects proved that the olfactory sense is temporarily weakened within 15 minutes of smoking. There is no other evidence to show that smoking interferes with the olfactory ability after this 15 minutes. Smokers actually detect the artificial musk smell known as galaxolide better than non-smokers. However, for non-smokers the reverse happens with the musky-urinous smell of androstenone. The skunky smell of mercaptan was given a higher-ranking pleasantness score by the smokers, but so did the rankings of rose and nutmeg. It seems that smokers become more sensitive to certain smells, while becoming less so to others. In conclusion, the minor effect of smoking observed in the clinical tests seemed to have no significant effect on the everyday ability of the olfactory senses. In fact, many perfumers who are the best in their business are heavy smokers. Another interesting fact is that while normal olfactory ability deteriorates with age, those of perfumers gets better with age, as decreased sensitivity is compensated by their experiences and techniques.

The important factor is harmony

Normally, the individual odorous chemicals that form compound flavors give many impressions, such as a very strong feeling, a reminder of something, or a resemblance to something else. However, often such chemicals do not feel like good odors. With the exception of lemon, orange, and some flowers, not many odors are refreshing or pleasant. Nevertheless, once they are combined to make a fruit flavor or a perfume, they smell really good. There are many odorous chemicals that smell bad individually, but have surprising effects in odors once they are appropriately combined together. Indole is probably the most unfortunate molecule because it smells really bad in high concentrations, but once diluted it is an obligatory chemical for lily and tuberose fragrances.

Food, excluding freshly served salads or fruits, also tastes better after being prepared by a chef. Without cooking, food does not taste good. Flavor compounding is very similar to cooking. For example, a spicy black pepper alone is electrifying in your mouth after a bite. On the other hand, combined with some sauces, the pepper has the power to create an exquisite taste. We must know which smells should be mixed in which quantity ratios in order to bring out the unique characteristics of individual fragrant ingredients. The most important principle in flavor compounding is balance, and thus the objective is to find the harmony between the odorous ingredients.

The first smell that comes out of a fragrance bottle is called a "top note." This is the first impression of the odor and is similar to the taste of food. It is very important whether this first impression is attractive or not. Thus, creating an attractive top note and figuring out how the odor disperses in the air are important issues for perfumers when compounding odors. After the top note has elapsed, the flow of a rich aroma follows. This is the "middle note." The last wave is called the "last note" or dry-out. In reality, these concepts are important for perfumes and fragrance products where the odor lasts for days. Yet for food products, it is not good for the odor to last a very long time. Only the top note and the general flavor are important for food products.

The balance of odors determines the future of a particular flavor compounding. To understand this balance, the perfumers and flavorists must identify the characteristics of the single odorants accurately, based on their knowledge related to the olfactory senses. If it is a natural odor, the olfactory knowledge may include its history, botanical species, origin of production, extraction method, physical properties, the proportions of the major ingredients, price, and areas of application. For synthetic odorants, it refers to the raw chemical materials, the production method, physico-chemical properties, price, and applications. In addition, they must know about the consistency, stability, safety, solubility, and purity of the odors in various solvents.

Furthermore, countless experiments on combinations of different odorous chemicals must be conducted. For example, in order to create the beautiful red-purple color of a morning glory, painters must conduct numerous experiments to find the right paint colors and the correct mixing ratio. These harmony studies should build upon each other to obtain an experience-based knowledge of compounding, and to find a good harmony, trying out various combinations is the only solution. Therefore, it is important for perfumers and flavorists to experience a wide spectrum of primary odors and actually use various aromas to become familiar with the perception of the fragrance products.

Applications of compound flavors

A compound flavor is a simple mix of odorous chemicals. Once a perfumer has prescribed the best combination of natural or artificial odorous materials, the manufacturing plants follow the same recipes to produce the perfume products. Figure 6.1 describes the general industrial processes of compound flavors. Perfumery ingredients are less restricted and are often higher in concentration than those used in food products, but the compounding principle is the same in both industries. Food flavors are made by diluting an odorous material in food-grade solvents, such as water, ethanol, or oils, to a final concentration of 0.1% in the food products. Depending on the preferred forms of the final flavor products, they are processed through emulsification or powdering.

Types of odorants

Alcoholic water-soluble odorants: These are usually called essences. They are extracted from odorous materials by hydrated alcohol. They dissolve in water and are volatile, with low heat stability, which is generally disadvantageous, but can also be an advantage in some applications. These odorants are used in various fields, including fermented milk beverages, yogurt drinks, ice cream, and confectionary, because they can create a light, fresh, and elegant scent. They must be used in low-temperature applications because they are sensitive to heat.

Oil-soluble odorants: These are created by dissolving the aroma base materials in vegetable oils. They are not soluble in water and have

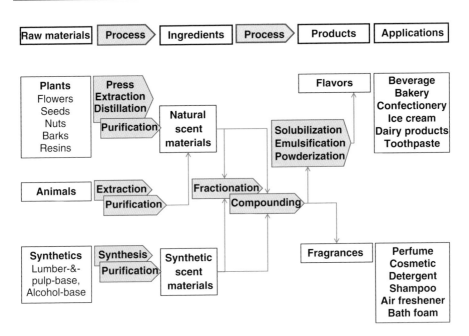

Figure 6.1   Diagram for the manufacturing processes of compound aromas for the flavor and fragrance industry.

high heat resistance. They are used in the baking industry, where high temperature operations are commonly used in the manufacturing processes.

Emulsified odorants: An oil-soluble aroma base is emulsified in water using emulsifiers, stabilizers, or special polysaccharides. They disperse in water and give a unique cloudiness. Their purpose is to give flavor and dispersion stability to juices and beverages.

Powdered odorants: These are produced by spray drying the emulsified aroma base with the coating materials. Because the aroma base is coated with polymeric coating materials, it must have a high stability to maintain the quality of the product. They are used in powdered drinks, powdered soups, cake mixes, chewing gums, and seasonings.

Synthetic flavors versus natural flavors: which is safer?

There is a misconception that all natural food ingredients are safe and all synthetic ingredients are not. Similarly, people worry if an odorous chemical is synthetic. Even if the synthetic odorants are chemically the same as the natural chemicals, people may have prejudices against the synthetic odorous materials. They think casein (milk protein) is a nutrient, but consider sodium caseinate to be a dangerous synthetic chemical. Some people also think glutamic acid is a useful amino acid, but consider that sodium glutamate is an artificial flavor enhancer. These misconceptions have not been shown to be correct or proved by any scientific evidence, but are propagated by articles and opinions in the mass media. A natural odorant has a much more complex composition than an artificially created odorant. Over 1000 different odorous chemicals have been identified in coffee alone, 400 in tomatoes, and hundreds of chemicals in bread and green tea. Compound flavors (artificial flavors) are chemically the same as natural flavors (simply with different origins). These compound flavors contain only selected odorous chemicals at concentrations the same as those of natural flavors. Therefore, there are smaller amounts and fewer types of odorous chemicals in artificial flavors than in the natural flavors, but they are sufficient to exhibit the same effectiveness as the flavoring ingredients.

The belief that anything natural is safe may relate to ignorance about the history of life and evolution. All of the toxic chemicals on earth were made by plants and microorganisms for self-defense. Everything in this world is a chemical substance. They all have the same characteristics, if their chemical structures are identical, regardless of how they were made, whether by a plant through natural biosynthesis or by a chemist in a laboratory or in a production facility. Once in a while there are some exceptional cases when isomers and impurities are not completely identified, but, nowadays, all approved artificial chemicals have been proven to be safer than naturally made chemicals. The reason why natural odorous chemicals are also safe is not because they do not have any dangerous components, but because the level of dangerous components is very minimal in the plants that are used. On the other hand, the use of any dangerous components is forbidden in artificial additives by regulations for food, drug, and cosmetic products. Therefore, all approved artificial additives are safe.

A flavorist imitates a complex natural flavor with 10–30 odorous chemicals from the thousands of naturally existing odorous chemicals in foods (Figure 6.2). Artificial flavors are produced to be of a very high purity through strict product quality management; thus, they are very expensive. Since it is not very economical to use a lot of these high-end artificial ingredients, a flavorist conducts hundreds of experiments to reinvent the natural flavors with the least amounts of

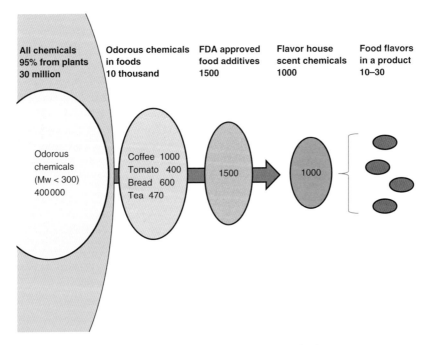

| All chemicals 95% from plants 30 million | Odorous chemicals in foods 10 thousand | FDA approved food additives 1500 | Flavor house scent chemicals 1000 | Food flavors in a product 10–30 |

Odorous chemicals (Mw < 300) 400 000

Coffee 1000
Tomato 400
Bread 600
Tea 470

1500

1000

Figure 6.2   Types of odorants out of all the odorous chemicals.

artificial ingredients. Even ignoring the costs, it is impossible to create the exact same flavors using all the same chemical ingredients existing in the natural flavors, because more than 80% of naturally existing odorous chemicals are not approved for food, drug, or cosmetic products. This is due to safety regulations. In fact, in coffee, the pure chemical forms of many of the odorous chemicals have been identified as possible carcinogenic substances.

Stanford University Chemistry Professor, James Collman (2001) wrote that:

> To some degree, almost everything has the potential to be toxic, even those substances we require for our life, such as oxygen. Nothing is absolutely safe; both natural and artificial chemicals can be dangerous. The safety and/or effectiveness of any particular substance varies with its concentration and with which part of the body is exposed to the substance. When tested in high doses on animals, a half of approved chemicals for food, drug, or cosmetic products turn out to be carcinogens, whether they are natural or artificial chemicals. They do not work carcinogenically under regular conditions because we do not consume the chemicals like the animal experiments. When we consume them in

small amounts, in many cases, they are harmless. Plants develop their own insecticides. Of the 64 natural plant insecticides that were tested, 35 are carcinogens. For example, more than 1,000 chemicals have been identified from roasted coffee; 26 cancer suspected substances have been tested, and 19 of them were found to be carcinogens. Each day, Americans eat 1.5 gram of such natural pesticides on average; this amount is more than 10,000 times greater than the ingested amount of residual chemical pesticides. Forty-three foods were found to contain at least 10 ppm of natural carcinogenic chemicals. Among them such as parsley, oregano, sage, rosemary, and thyme are included.

James Collman (2001)

Nutmeg, which has had a great influence on the history of medieval Europe, has been used as a medicine. In China, it was used to treat rheumatism and stomach pain, while in Southeast Asia it was used to treat diarrhea and stomachache. In Europe, not only was it used as an aphrodisiac and an anesthetic, but it was also used as a preventive medicine for the Black Death. It is possible that isoeugenol molecules from nutmeg kept away the fleas that carried the plague bacteria. Nutmeg also contains insecticidal chemicals, such as myristicin and elemicin. However, it was safe for consumption because in medieval times nutmeg was very rare, and the insecticidal ingredients would only be available in small quantities. One whole nutmeg will make a sensitive person feel sick and sweat a lot. It will also accelerate the heartbeat as well as increase blood pressure and sometimes the person may also hallucinate for a few days. Consuming more than one nutmeg can be fatal. Moreover, consuming a large quantity of myristicin can critically damage the liver. There are many kinds of natural chemicals that can be toxic and fatal to the human body, including poisonous mushrooms and solanine in potatoes. Oxalate in spinach, hydrazine in mushrooms, and myristicin in carrots or celeries are also life-threatening toxins. There are some natural products that even contain formaldehyde. So the statement that a natural product is good and safe for the body is false, which is why we cannot randomly pick and eat wild mushrooms or berries on a mountain trail. Of course are poisonous chemicals in natural products that people have been eating for a long time, but they are okay to eat because the level of poisonous chemicals is insignificant. However, we cannot allow these chemicals to be used freely on the basis that they exist in natural products and historically have been consumed over time.

Apple and strawberry flavors are not made from the odorous materials extracted from apples and strawberries, but they are actually made with the same odorous chemicals that affect the total flavor characteristics of apples and strawberries most effectively. In other words, only the source of the effective materials is different. Besides, the effective odorous materials are the only components that contribute to the actual aroma of apples and strawberries. So, the types and quantities of odorous chemicals used are less than the natural flavors in real apples and strawberries. All the ingredients of natural flavors are not considered to be poisonous because they are found in natural apples and strawberries. However, many of them demonstrate toxicity when they are present in very high quantities. Nevertheless, only the substances that are proven to be safe to use are chemically synthesized as ingredients for artificial flavors. In artificial flavors, there are no chemical substances for which the toxicity has not been identified or which are unknowns. Thus, it is meaningless to ask whether artificial flavors are safe or not. In the same way, it is also meaningless to question which flavors are the safest, artificial flavors or natural flavors? Even a natural flavor that is made from hundreds of unknown chemicals is considered safe at the level of normal consumption; so how much safer can it be than an artificial compound flavor that is made from less than 30 types of scientifically proven safe chemicals? This is a very reasonable answer to the question. Thus artificial flavors are safer than or equally as safe as natural flavors.

A flavorist comes into contact with more chemical substances than other food researchers and is surrounded by 1000 times more chemical substances than a normal person. Once, a painter came into the laboratory at my flavor company to do some interior painting work. He complained many times about the smell of all the experimental odorants in the laboratory. However, this was an awkward situation as a flavorist in the laboratory also had a hard time working with the paint odor.

Over time the amounts of odorous chemicals experienced by normal people are far less than those sensed by a flavorist. Some famous foreign flavorists have continued working well past the normal age of retirement. So is the position of a flavorist dangerous or safe throughout their working life? If any odorant used by normal people was dangerous, the flavorists would not be healthy, and if they did not remain healthy, it would be impossible for the flavor industry and perfumeries to be sustained as legitimate industries.

The press have spoken of the dangers of artificial butter flavor (i.e., diacetyl) after a worker at a flavor plant developed a health problem with his lungs. However, this is not an evidence-based statement. Diacetyl exists in all dairy products. It is found in everything from milk, butter, and cheese, and it is the substance that provides the milk flavor to these dairy products. The lesson from this lung patient is that constant long-term exposure to a high concentration of a natural chemical can be dangerous; therefore, specific controls are required to protect a worker's health. However, instead the press threatened that consumers should be careful because artificial flavors might contain diacetyl. Another example is alcohol; microorganisms cannot grow in 10% alcohol. Indeed, solutions with 75% alcohol are very effective sterilizers. However, we enjoy drinking such threatening substances at various concentrations: wine has 9–16%, soju has 18–45%, vodka has 35–50%, whiskey has 40–60%, and rum has 38–80% alcohol. Alcohol is a leading carcinogenic substance, but some argue that small quantities are good for our health. Likewise, the only, but critical, difference between a toxin and a medicine is the matter of quantity. But the press are only interested in provocative reports. They are unconcerned about the fact that this lung patient issue was an isolated incident that only occurred once in the whole world, and that special care is given to ventilation systems, for the health of the workers, so that such a problem does not arise again.

Tar-based food additives and colorants have not been fairly evaluated with respect to their safety. Some people believe that most flavors are dangerous just because they are made from tar. In fact, tar has a long history: in 1792, large quantities of coal tar were produced as a byproduct of coal gas manufacturing; in 1845, benzene was successfully extracted from coal tar distillate; and then in 1856, Sir William Henry Perkin synthesized an artificial dye that completely changed the dye and fabric industry. During the Easter break, Perkin decided to synthesize quinine, the miraculous medicine for curing malaria, in a small laboratory installed in his house, after hearing his mentor (August Wilhelm von Hofmann) saying that quinine can be synthesized from phenol (or carbolic acid), a particular chemical in coal tar. At the time, quinine was the only preventive medicine and treatment drug for malaria. Quinine is obtained from the bark of a *Cinchona* tree in South America, but after excessive harvesting, the *Cinchona* tree became an endangered species. However, Perkin's experiments did not succeed. As a result of his experiments, he had made a black colored chemical, which he dissolved in ethanol to give a dark velvety purple solution. He then submerged some silk fabric pieces in this purple solution, and they were dyed this dark velvety purple color. This purple "mauve" dye was subsequently developed by Perkin, and it swept through the fashion industry in 1859. After the successful synthesis of mauve, many similar processes were developed to produce various other artificial dyes from coal tar. This dye industry also became the parent business for the organic chemistry industry (antibiotics, explosives, perfumes, paints, inks, pesticides, plastics, etc.).

However, it was in Germany that organic chemistry grew through massive industrial applications rather than in England or France. Even though mauve was first developed in England, and the dye and fabric industry were treated as important businesses for a couple of centuries in France, they did not realize the usefulness of coal tar and the importance of the organic chemistry industry, which resulted in benefiting Germany instead. BASF and Bayer, great chemical companies of Germany, earned much of their profits through their dye business. Bayer might remind you of Aspirin, but when they were established in 1861, they originally produced a dye called aniline. Aspirin was synthesized in 1853, but was not offered on the market until around 1900 after Bayer had invested their profits from synthetic dyes (especially alizarin) into pharmaceutical products in a major effort to diversify their business. At that time, Germany monopolized 90% of the global synthetic dye market. Dyes changed the history of human culture. Within a century of the synthesis of mauve by Perkin, giant chemical companies dominated the dye market and the newly emerging organic chemistry industry. Today's antibiotics, pain-relievers, and other pharmaceutical products are a result of the profits and knowledge earned by the chemical companies of that era. Tar itself is a medicine. Before steroids were commercially available, for over a century tar had been used as a medicine to treat skin troubles, which extended the usefulness of the byproducts of coal distillation. The stickiness of tar helped the efficiency of its medicinal effect. Aspirin was also chemically synthesized from tar.

Anyone who worries about the safety of tar because of its source must remember the origins of petroleum. Petroleum is made from ancient organisms, and is an untouched and purely natural matter without any interference from humans. Even though the natural matter (crude oil) is separated into various fractions by their boiling temperature differences, this is not natural matter being converted into synthetic material, in the same way as coal tar is a purely natural material. The reason why many organic chemical products are made from petroleum is because it is the cheapest organic material (natural matter), not because it has some special powers. Even if chemical companies produce various chemical products from other natural matter that is considered more natural and environmentally friendly, the resulting chemical products would still be the same as the materials from petroleum. The only difference would be the higher price due to the more expensive raw material costs.

> Matter is matter. It is not honorable or foul. It is infinitely changeable. Its origin is not important at all.
>
> *Primo Levi (1919–1987), an Italian Jewish chemist and writer*

All odorous materials are regulated by government authorities or industrial associations. South Korea is probably one of the few countries where the government regulates all flavors. Since 2005, all substances that can be used for flavors and fragrances have been regulated and only these approved odorous chemicals or extracts should be used for manufacturing flavors and fragrances. It is the most transparently administrated system in the world. In the United States, the Flavor and Extract Manufacturers Associations (FEMA) oversee the flavor industry. In Europe, flavors are excluded from additives. In Japan, the government approves the use of only around 80 types of fragrant chemicals and leaves their specific applications to the industry. For the case of aromas, these are mostly natural extracts or synthetic chemicals with the same structures as the natural extracts. The amounts of chemicals used can be detected by smelling the product. However, a preference for a product does not increase with the level of the chemicals, this only increases the cost. Because of this preference for odor intensity and the related costs, the usage of odorous chemicals is automatically limited. Therefore, government authorities could pass the dosage control over to industry.

There are many odorous chemicals, but only a fraction of them are used regularly. For example, in Japan, 15 odorous materials account for 64.5% of all the odorants used. The number of items that are of practical use to the industry is not large, and most of these were developed before 1980. Japan is unrivaled in Asia's flavor market but for most odorous materials less than 1 kg is produced annually in the country. From these statistics, it is easy to understand that the total amount of odorous chemicals in an individual consumer product seems very low. Odorous materials have a molecular weight of less than 300 (an average of 200), meaning that the molecular weight is too small to exhibit a specific medicinal function or to be accumulated in the body. They are neutral organic chemicals easily excreted from the body after providing only sensory characteristics.

Advantages and limitations of natural flavors

Most natural flavors are well balanced, or to be precise, they are the most familiar fragrances. Not all natural flavors are expensive. Orange oil is the cheapest natural flavor. From 1000 kg of oranges, the orange industry produces 100 kg of orange juice with 65% sugar content, 3 kg of orange oil, and 1 kg of orange flavors from the distillate. There is no cheaper odorous chemical than the orange flavor. It would cost more to synthesize the orange flavor chemically. From this orange oil, removal of a part of oil soluble fractions using alcohol produces orange essence. This water or alcohol soluble orange essence is used for various purposes. Every citrus flavor is made from natural fruits.

Nevertheless, it is better when fruits are consumed as fresh fruits. With the exception of few items, it is undesirable to extract the flavor chemicals from fresh fruits when the flavors exist in the fruits at extremely low levels. It is always good to understand the advantages and disadvantages of natural flavors.

- The amount of odorous chemicals in every plant is less than 0.1% except for citrus peels. Thus, because the purity and concentration of natural flavors is low, for the purpose of creating the desired effect, the throughput of the naturally extracted flavors must be very high in a product. Therefore, the product containing a high level of flavor extracts is frequently unsatisfactory and unstable with respect to physical and textural properties.
- In most cases, a concentration process is required to increase the fraction of the odorous chemicals for this to be practical through extraction or distillation. These processes are usually accompanied with substantial changes to the quality of the flavors.
- There are significant deviations in the intensity and quality of the natural flavors depending on the source of the product, the ripeness at harvest, and the conditions of post-harvest handling.
- Supply of natural flavors is becoming uncertain with increasing regulatory restrictions due to environmental issues.
- Most natural flavors are unstable and change significantly during post-harvest handling, processing, or storage.
- Most natural products have an enzymatic system that gives off an off-flavor or decreases the intensity of the flavor.

- For many natural products, their evidence-based toxicology information is not clearly identified. If they were investigated accurately, like synthetic chemical compounds are, the toxicology results would be quite surprising.

Advantages and limitations of compound flavors

There are many advantages of compound flavors and how they are used:

- They have higher purity relative to natural flavors. Thus, they are used in smaller quantities, which reduces the cost and makes the management of the supply chain less dependent on price changes.
- They are safer and have a longer storage period. It is also possible to create compound flavors that are more stable even under extreme processing conditions.
- Owing to the high concentration and purity, they can be processed in multiple forms (e.g., spray-dried powders, coatings, capsules) to fit various specific applications.
- They can be procured in a timely manner because there are no seasonal restrictions, such as harvest time, or other limitations to their supply, which is also environmentally friendly.
- Production specific to custom orders is possible. This allows the final consumer products to be differentiated from other products.
- Undesired properties or fractions in natural flavors can be adjusted or eliminated.
- Quality is consistent.

Natural flavors, not always but in most cases, are more delicate and full. Fruits have less variation in their flavors because their composition is simple. However, it is very complicated to describe all other flavors that are created through fermentation or thermal processing, such as coffee and vanilla. Thus, it is difficult to recreate the complicated original flavors by compounding over 400 flavors with only 20–30 odorous chemicals. Many odorous chemicals in natural flavors contain significant amounts of their precursors, which may exhibit extra odors under particular conditions, but compound flavors with only 20–30 chemicals do not have precursors. In the past, when technology was less developed, compound flavors were evaluated as artificial and unbalanced. Fruit flavors composed of only a few odorous chemicals became outdated products in the market. Nowadays, only the compound flavors that are very similar to the real fruits are selected and utilized commercially.

Almost everything we eat and drink has flavors. It is scientifically incorrect to insist that the use of natural flavors is the best way to supplement the limited flavors in natural products. However, marketing strategies to advertise the advantages of natural flavors will remain popular for many more years. Nevertheless, it must be clear that synthetic odorous chemicals and natural odorants are the same molecules and they are neutral ingredients that only serve the sensory functions but have no physiological functions.

References

Castriota-Scanderbeg, A., Hagberg, G.E., Cerasa, A., Committeri, G., Galati, G., Patria, F., Pitzalis, S., Caltagirone, C., Frackowiak, R. (2005) The appreciation of wine by sommeliers: a functional magnetic resonance study of sensory integration. *Neuroimage.* **25**(2): 570–578.

Collman, J. (2001) *Naturally Dangerous: Surprising Facts about Food, Health, and the Environment.* University Science Books, Sausalito, CA. ISBN 978-1891389092.

Fidelity (2013) *Sectors & Industries Overview.* https://eresearch.fidelity.com/eresearch/markets_sectors/sectors/sectors_in_market.jhtml (accessed 17 July 2014).

Gilbert, A.N. (2008) *What the Nose Knows: The Science of Scent in Everyday Life.* Crown Publishers, New York. ISBN 978-1-4000-8234-6.

Kang, S.M. (2009) *Grand Perfumeur, Thierry Wasser.* Joongang Ilbo. Seoul, Korea. http://article.joins.com/news/article/article.asp?Total_ID=3939477 (accessed 17 July 2014).

USDA, United States Department of Agriculture (2012) *Ag and Food Statistics: Charting the Essentials.* http://www.ers.usda.gov/data-products/ag-and-food-statistics-charting-the-essentials.aspx#.UrI6r_RDu8Y (accessed 17 July 2014).

The Freedonia Group (2013) *Flavors & Fragrances - Demand and Sales Forecasts, Market Share, Market Size, Market Leaders.* The Freedonia Group, Inc. http://www.freedoniagroup.com/DocumentDetails.aspx?ReferrerId=FG-01&studyid=3044 (accessed 17 July 2014).

Tieman, D.M., McIntyre, L., Blandon-Ubeda, A., Bies, D., Odabasi, A., Rodriguez, G., van der Knaap, E., Taylor, M., Goulet, C., Magero, M.H., Snyder, D., Colquoun, T., Moskowitz, H., Sims, C., Clark, D., Bartoshuk, L., Klee, H. (2012) The chemical interactions underlying tomato flavor preferences. *Curr. Biol.* **22**: 1–5.

7 How Flavors Influence us

Brain development began with the olfactory sense

There are over 388 types of genes assigned to the olfactory sense. Out of all the 23 000 genes in the human body, this is the largest number of genes assigned to one particular function. This can be overwhelming compared with the number of genes related to other sensorial systems. For example, four genes are used for the visual sense (two in nocturnal animals and three in humans) of vertebrates. In fact, 50 different types of anosmia have been identified; this means that individual genes are used to distinguish smell through the combination of signals received by all the olfactory receptors. Even though olfactory genes have deteriorated to only 388 types, the importance of each gene remains constant. Also, smell is the only sensory system that we cannot randomly shut off. We can smell whilst sleeping as long as we are breathing. It is the oldest sensory system in biological organisms, and there are no other functional organs that have developed or are controlled by more than 388 genes.

The first mammal in history was very similar to today's porcupine. If hagfish is the primary example of all primitive vertebrates, the porcupine is the first example of all primitive mammals. Hagfish are almost blind, but have well-developed senses of touch and smell. The olfactory function takes up the largest area in a porcupine's brain. The olfactory cortex of a porcupine is similar in size to the prosencephalon (forebrain). The amygdala and hippocampus, which are the parts that cause emotion and ancillary reminiscence, are particularly noticeable. In general, a porcupine's brain is mostly taken up with the olfactory sense and associated organs, while areas occupied by the other senses are much smaller.

How Flavor Works: The Science of Taste and Aroma, First Edition.
Nak-Eon Choi and Jung H. Han.
© 2015 John Wiley & Sons, Ltd. Published 2015 by John Wiley & Sons, Ltd.

Mammals avoided contact with dinosaurs and, as a result, evolved as nocturnal animals. Thus, their senses of hearing, smell, and distance advanced rapidly, while the sense of vision developed slowly. After the extinction of the dinosaurs, certain evolutionary features, such as the expansion and refinement of vision steadily appeared. The neocortex, like the olfactory cortex, is in a random-access structure. Thus, the original point-to-point images disappear during several signal transduction stages in the visual cortex, producing only critical final image information. The visual system is an imitation of the olfactory system. Half of the brain of a butterfly is dedicated to the sense of smell. Not only did the olfactory system develop as the first notable sensory system, but the development of brain functions for other senses largely benefited from that of the sense of smell.

The target cortex area for the olfactory sense is in the frontal lobe, which has the most layers even within the cortex. The frontal lobe is located in the front of the cerebral cortex and extends to the spatial area between the nose and forehead. Originally, it was a structure to control motorization in small-brained mammals; it is connected to the corpus striatum and is responsible for movement and other complex motions. When the corpus striatum determines the muscle control structure of the brain and conducts several coordinated movements, the frontal lobe reviews them and begins to predict the result of these movements through learning mechanisms. After the frontal lobe begins to control the corpus striatum, it then decides which movements to make under certain circumstances and, as a result, achieves the ability to plan. In the brain, no structure functions alone or independently. They all send and receive particular forms of information. Specifically, both the hippocampus and amygdala send signals to the corpus striatum, which is then connected to the thalamo-cortical circuit. All structures come together to form a universally functioning structure. There is a theory that the system responsible for the sense of smell also became the basis for the brain of modern humans. Whether it is a good or bad emotion in the amygdala, it will influence the formation of memories in the corpus striatum structure and the thalamo-cortical circuit. Many functions that are necessary for brain function are embodied in the olfactory system. These functions include: adapting (olfactory fatigue) by controlling the signals from the olfactory cortex to the olfactory bulb, memorizing and learning by amplifying or reducing the loop signals according to a specific smell and eventually by strengthening the synapse connection, comparing the memorized pattern with a new input pattern, amplifying the desired signals and noise reduction, inferring the prediction of the total information from partial information by filling in gaps, and uniting to combine the sense of smell with other senses. Perceiving smell has a similar mechanism to facial recognition through the visual sense, and the olfactory circuit is similar to the hippocampal circuit.

In the past, through chasing after or escaping from various foods odors, animals, and the environment, and through their success or failure, the synaptic connections of the cerebral cortex and corpus striatum became modified each time. Thus, when facing the same situation again, the striatal-thalamo-cortical circuit, which has been built through hundreds or thousands of experiences, interacts with the stimulant. Through this program, composed of a "built on experience" circuit, the actions of organisms possessing brains can change from one decision sequence to another. Because cyclic loops are made from these modified synaptic structures, vast frontal striatal-thalamo-cortical circuits can expand all responding actions after circulating the signal for a long time (maybe for a couple of seconds); and the neurocircuits can memorize a modified current situation and decision loop. Such work of decision and memory is a necessary instrument in mammals for their brain development. Thus, the frontal lobe, that is, the CEO of the brain, can plan a big picture to achieve its objective and organize work to breakdown the structures of detailed activities. First, the frontal lobe is stimulated by a new olfactory signal; then, the cerebral cortex is stimulated to select the next action. These steps continue until the perceived odor settles in the brain with full information connections. Through this process, the animal acts intelligently and naturally.

Advances in neuroscience are considerable, but there are more unknown areas than those already discovered. There are one million neurons packed around the radius of one neuron in the brain. Each neuron forms synapses with 10 000 neurons, which is only 1% of all possible synapses. Therefore, the connections with other neurons are always changing. Chaotic disorder, rather than order, is the basic identity of the brain. This chaos holds in place the balance of contrasting objectives, such as flexibility and stability, and adaptability and autonomy.

The sense of smell is the dominant sensory system in animals. It is also a starting point for understanding the brain function. The mechanism of directivity first appeared in the olfactory system. In contrast to the visual and hearing systems, which require a significant level of preparatory processes for precise perception, the sense of smell is simpler. Anatomically, the olfactory sensory system in the nasal cavity is positioned closer to the limbic system and directly touches the part of the brain that is involved in emotional expression. Brain functions cannot be properly understood without a good understanding of the sense of smell. Therefore, neuroscientists must research olfactory senses more extensively whether they like it or not. Fortunately, this helps the food industry to understand the functions and mechanisms of flavors more accurately.

The human olfactory sense is less sensitive and inarticulate

A dog is 2000 times more sensitive to the smell of the chemical thiol and 10 million times more sensitive to butyric acid than humans. It is not because the olfactory organs of animals perform better than a human. It is just that they have more olfactory cells. While humans have 10 million olfactory cells, a rabbit has 100 million and a dog has 1 billion. Even a small mouse has more olfactory cells than a human. The reason why each animal has a different olfactory ability is because the importance of the olfactory sense for their survival strategy is different for each animal. Because birds fly high above the surface, their sense of smell is not very important to their survival. Thus, their sense of smell has deteriorated, while their vision has developed very well. The most sensitive sense of smell probably belongs to four-legged animals. A pig can smell a truffle 15 cm under the ground. Snakes and insects also have a heightened sense of smell.

The cerebral center of the olfactory system is a part of the limbic system, which is in charge of emotion and memory. Since emotions are not as precise as logic, the sense of smell, which is closely related to emotions, is not as precise as vision or hearing. Also, the sense of smell is not really related to the language center in the left hemisphere of the brain; thus, it is hard to describe smells verbally. Visual stimulation can be described in detail, such as azure or rusty yellow, but smell cannot be described as effectively. Sure, expressions such as the smell of stew or bacon are used because of the clearly distinctive smells of stew and bacon, but it is difficult to actually describe the feeling with words when simple smells are mixed. Smells cannot be communicated with words. Moreover, the sense of smell is slow compared with vision. If a blindfolded person is asked to smell a sample, and is then asked to smell it again an hour later, the person recognizes the second smell as the same as the previous smell after 12 seconds. Approximately 20% of the people say that the second sample is different from the first sample. As such, the olfactory system is a slightly deficient and slow sensory system.

Humans' sense of smell has degenerated greatly

The cerebral center of the olfactory system is only 0.1% of the whole brain. It is a very small proportion considering that the olfactory part is one half of the brain of a butterfly, and a third of the brain of a rat. The degree of degeneration of the human sense of smell is easily estimated when we understand that there are only 388 functional olfactory receptors in humans out of 1000 (Figure 7.1). This means that over half have deteriorated. The number of fossilized receptors is 18% in mice, in lemurs of Madagascar, and in New World monkeys in Central and South America, whose senses of color are unreliable. In black-and-white colobus monkeys in Africa and in Old World monkeys in Africa and Asia, 29% of their receptors have fossilized. In anthropoids (orangutans, chimpanzees, and gorillas) other than humans, 33% of their receptors have fossilized. This shows that in relation to other anthropoids, the human sense of smell has degenerated much more.

However, it is a mistake to think that the human sense of smell is insignificant because it is not as sensitive as those of other animals. Degeneration is also a part of evolution. The eye has a very complex sensing system that has taken from 500 000 to 1 million years to complete through evolution. As the visual sense became unnecessary inside a cave, the eye completely degenerated over 1000 years for cave animals. Evolution is the process of creating necessary functions as

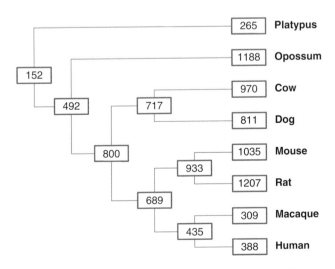

Figure 7.1   The change in the number of olfactory genes of different animals over the last 200 million years. Modified from Nei *et al.* (2008).

well as eliminating unnecessary functions. The evolutionary strategy is to decrease numbers but to enhance special functions. Humans use highly developed brains. Brain development means an increase in the ability to integrate functions, not an increase in sensorial ability. The olfactory parts in human brains became relatively smaller in size because the human brain grew bigger, not because the absolute size of the olfactory parts was smaller. The whole area of the human olfactory sense is bigger than the size of other small animals.

Proust phenomenon: odor-evoked autobiographical memory

The sense of smell is directly linked to learning, memory, and emotions. The hippocampus and amygdala are involved in the most fundamental vital activities of humans. The memory of an odor that brings up strong emotions can last quite a long time. Relatively speaking, the cerebral center for languages in the brain was developed much later than the area for olfactory sense. A memory described with language is much more visual and rational, but it cannot keep up with the abundance of emotions derived by the memory of a smell. Memory based on language is hard to maintain without the aid of record keeping, but memory based on smells can be evoked with the smallest suggestion at anytime and anywhere. There is scientific evidence that olfactory stimuli can cue autobiographical memories more effectively than cues from other sensory modalities. Explanations for these effects can be invoked from accepted principles in contemporary cognitive psychology (Chu and Downes, 2000). Stimulating odor takes us through time and space by igniting both the olfactory area and the memory area of the brain. An olfactory simulation makes a memory pop up as if a landmine long hidden in the bushes suddenly explodes. It is as if the smell nudges the detonator, and all the memories burst out at once. The smell of burning leaves does not only bring up the memory of eating baked sweet potatoes or barbequed meats, instead, emotions and memories of old family parties are more vivid than the smell. Because the effects of smell appear instantly, they stimulate one's emotions before one even has a chance to think about the smell. For this reason, when we perceive a certain smell, we are carried away by unknown emotions linked to the smell before even beginning to understand the nature or cause of the emotion.

Emotion also plays an important role in olfaction ability in reverse as it becomes more sensitive when something bad happens. This makes sense if we imagine our ancestors smelling an unusual odor right before encountering a ferocious animal. The olfactory sense becomes much more sensitive when we are in danger. The brain has developed through evolving the sense of smell. The neocortex, which controls the human characteristics involving vision and language, evolved later after the olfactory cortex was inflated. Although vision collects the majority of information we perceive and, therefore, is known as the most humane and high-dimensional sense, the foundation of the visual sense was set by the olfactory sense.

Sensorial preference changes destinies

The panda was originally omnivorous. About 4 million years ago, its umami taste receptor genes were broken down. The panda does not have a umami taste receptor and it cannot taste meat. Today it only eats bamboo leaves. In fact, a giant panda (*Ailuropoda melanoleuca*) has a chin, teeth arrangement, and digestive system made for meat consumption. The bear family (*Ursidae*), which includes pandas, enjoy fruits, and eat meats as well. Researchers analyzed fossils of 7 million year old panda teeth and found that pandas have been eating bamboo since that time. Dependence on a herbivorous diet grew as it became harder to obtain meats; pandas were finally transformed completely when they lost the umami taste receptor genes. Meats are about 1% carbohydrates; this means they are not sweet and all meat tastes of umami. If one cannot feel umami, then meats are not tasty at all.

In contrast, in 2005, it was revealed that the sweetness receptor genes had broken down in the cat family (*Felidae*), for example, in tigers. They do not have sweetness taste receptors (Li *et al.*, 2006). Thus they do not think about eating fruits because they do not actually know the rich taste of sweetness. In taxonomy, the cat family belongs to the *Canivora* Order, along with dogs and bears. At some point, one side lost sweetness and became carnivores, while dogs and bears remained omnivorous. Bears love honey, but to tigers, it is just a sticky liquid.

Among the herbivores, there are some animals that eat particular leaves or specific parts of plants. The koala is one of the large size animals among these. No other animal desires their food because eucalyptus leaves have a highly poisonous chemical, that is, eucalyptol. The koala can survive by consuming eucalyptus leaves alone. This is why koalas are known for being lazy and sleeping so much because they have no competition for their food source. Herbivores must eat a lot of plants, and, naturally, there are many poisonous plants. Herbivores, particularly ruminants, are most sensitive to the bitter taste because they have a higher chance of eating poisonous plants. Sweetness and umami taste receptors in humans are, of course, intact. This is why we are omnivores, and eat anything from grass roots to shark's fins.

Do silkworms only eat mulberry leaves?

Do herbivores only eat plants, and carnivores only meat? The answer is not necessarily. For example, the hippopotamus sometimes eats antelope meat and both carnivores and herbivores eat dirt sometimes. Ancient Native American shamans did not ignore this phenomenon. They discovered rock salt areas where carnivores regularly visited and licked something. They also discovered antiparasitics, antibiotics, and sometimes narcotic compounds by eating specific plants that carnivorous animals consumed only when they were sick. They thought some plants were poisonous when the plants induced vomit and diarrhea. However, later they realized that animals purposely eat these plants to empty their stomachs when they are sick.

Carnivores such as tigers and lions clearly eat plants during the summer, while herbivores eat feedstuffs that contain bone or meat powder (which is how the mad cow disease crisis started). The composition of mammalian milk is almost the same for herbivores and carnivores, except that the protein content of carnivore milk is slightly higher than that of a herbivore. Both carnivorous and herbivorous mothers eat the placenta of their offspring, a piece of meat. Rabbits and mice eat their offspring when they are under stress. Chimpanzees and baboons keep a herbivorous diet, but sometimes hunt. Gorillas, which seem to be carnivores, are complete herbivores. Birds are usually split between carnivores and herbivores, but their eating habits can change completely depending on their environment. The reason why they decide not to eat certain foods even though they can access this food is because of the food's lack of sensorial pleasure.

Silkworms, which insist on eating only mulberry leaves, could eat other things if their taste receptor DNA was changed. If every caterpillar could eat every leaf, a competitive food war would take place between different species. In the world of insects, food preference distinction decreases territorial conflict and food competition. However, they will eat anything if it is necessary. Pandas will eat meat if there is nothing else to eat. Therefore, instead of stating "sensorial preference decides destiny," it would be better to state "destiny depends on the senses."

Humans live with smells

Stage of development

Fetus: The first sense that a fetus can feel inside the womb is the sense of smell. A fetus can memorize some smells from foods that a mother has eaten and in the amniotic fluid in the mother's body. Spices that the mother enjoys during her pregnancy are preferred by the baby over other spices. If the mother drinks carrot juice during her pregnancy, the baby will like carrot juice. Thus, if someone wants their children to enjoy drinking carrot juice, it would be a very useful to keep this in mind.

Infant: The sense of smell is already developed in a baby who cannot even open his/her eyes. The baby identifies who its mother is through the olfactory sense. The baby smells food flavors from the mother's breast milk, which come from what the mother ate. The olfactory ability becomes more precise as they grow.

Child: Children enjoy sweetness, sourness, and saltiness, while strongly disliking bitterness. This is why children do not like vegetables in particular because they have less positive tastes (i.e., sweetness) and more negative tastes (i.e., bitterness). Children have a tendency to imitate the appetite of brothers, sisters, and friends more than their parents, which is a very independent phenomenon to the genetic preference inherited from their parents. However, compared with the fast development of the sensorial perceptions, their ability to suppress their brain's desires is established more slowly. This is the period of development that requires the most appropriate advice and training.

Adult: Appetite preferences and habits built during childhood are likely to remain throughout the whole of adulthood. Preferred appetites refer to the taste of a mother's food. The unique taste from her cooking has been passed on from her mother's cooking and so forth. Preferred appetites do change, but they are imprinted in the mind (i.e., memorized in their brain) even at an elderly age. As a person ages, the desire to find a memorized preference does not diminish. Families share not only DNA but also food as well. Thus, during the holiday seasons, they gather at one place and prepare food together to share their communal preferences.

Friends and people from the same hometown tend to have a similar food preference. To share the memories with others, people usually talk about old stories from their hometown and about common acquaintances, but the most important thing that unites them is sharing hometown foods. When they have hometown foods together, people from the same place are reunited. Therefore, restaurants that sell local foods have many old customers who do not search for new or exceptional tastes but stick with the same tastes experienced from their childhood and that are memorized in their brains. However, these instances are becoming less frequent because the tastes of commercially available foods are converging uniformly worldwide.

Elderly: One of the initial symptoms of Alzheimer's or Parkinson's disease is the loss of the olfactory ability. Patients say that the loss of smell causes a great feeling of loss and depression, and these feelings are not good for a person's health. Diminishment of the sensory ability adversely influences not only mental health but also physical health as well. Our gustatory and olfactory senses are also used to detect dangers such as fire, spoiled foods, and saltiness in food. Thus if someone loses the ability to smell, these functions also become very difficult to deal with.

What happens if you can no longer feel taste or smell?

If we lose our sense of taste and smell, we will not experience the taste of delicious food and, therefore, will lose our appetite. We will lose a large and essential pleasure from our lives. In fact, many elderly people live with this problem. The loss of the olfactory sense is not only a problem to the vitality of human lives, but is also a dangerous threat to safety and health because it makes the person unaware of fire, toxic gas, and spoiled foods. In the United States, more than 200 000 people visit their doctors every year due to olfactory sensory problems. In reality, many more people are suffering from sensory problems having lost their sense of smell and/or taste. Losing the ability to smell can cause emotional distress and isolation from society. Even losing the ability to smell unpleasant yet familiar odors can cause depression.

This can be demonstrated by the case of Dr Karl L. Wuensch in the Department of Psychology at East Carolina University who was eating food after he had heated it in the microwave. After eating a couple of mouthfuls, his wife came running in and screamed:

"How can you eat such smelly and spoiled food?"

If he had eaten just a little more, he would have become very sick. Sometime later, when Wuensch was gardening in his backyard, he sensed heat around the propane gas regulator. When he called his wife and son, they instantly smelled the leaking gas as they stepped into the garden. If his family had not been at home at the time, Wuensch could have lost his life. After these incidents, he realized that he had anosmia. He had lost his sense of smell. As a scientist, Wuensch not only then rigorously studied this disease, but also researched ways to help anosmia patients. He subsequently installed a propane gas detector at his home and was particularly careful not to let his food burn while cooking. He also learned the ways to reinforce the pleasure of eating even though he had lost his sense of smell.

Nearly 80% of flavors come from smells in the nasal cavity; the rest are from tastes in the mouth. Anosmia patients reinforce the remaining 20% by adding more spices or experiencing the texture of the food. In Wuensch's case, chili powder helped.

"I learned how to enjoy very spicy food. It is disappointing that my wife and son cannot join me in enjoying spicy food together."

Fortunately, Wuensch later experienced some success from treatment of his anosmia. At least it remained that way while he received treatment. He lost and recovered his sense of smell four times, and through this experience, he gained a unique perspective on the olfactory sense.

"After steroid treatment, I ran home right away and opened all the drawers and cabinets in the kitchen. I took out everything that had a smell. Crackers, cookies, and condiments, I took out everything that had a unique smell. I also ran out to the garden to smell the grass and rubbed a fistful of leaves on my face. I really felt pure happiness."

When Wuensch was asked which smell he missed the most after losing his sense of smell, he answered:

"The smell of people. I have never known I could miss the smell of people this much. I know a woman who fell into a deep depression when she couldn't smell the hair of her children. However, I never thought that anosmia could influence my social life so much. When I meet someone, whether very close or who I just know by face, I feel something is always missing. I knew before anosmia that everyone had their own smell. It could be cosmetics or their own body odor. Yet, once I couldn't smell it, that person seemed like a different person."

According to research articles published by two teams at the University of Dresden (Negoias *et al.*, 2010) and New York University (Hardy *et al.*, 2012), people with a weak sense of smell tend to be anti-social and easily become depressed. The research teams reached this result by asking adult patients about olfactory defects, daily activities, social relationships, favorite foods, and other aspects of their lifestyle. According to the interviews, one of the female subjects sometimes felt that she could not maintain a relationship with other people without being able to share the same sense of smell. The research teams showed that smell provides social information to others, and, thus, having a problem with smell means closing down a communication channel. They further added that this is the reason why people with olfactory problems have sex about half as frequently as a person with normal olfaction. They said:

"From these reasons, people who worry about their own body odor are more likely to avoid eating with others and making social connections."

Previous research showed that 1 in 5 people have an olfactory problem, while 1 in 5000 are born without any sense of smell.

Are humans free from pheromones

Humans only have remnants of the vomeronasal organ, which is known to be used to detect pheromones (Figure 7.2). The vomeronasal organ is a part of the olfactory organs of many amphibians, reptiles, and mammals. In humans, the vomeronasal organ exists in the nasal cavity of the fetus but deteriorates soon after birth. Since this organ is for pheromone detection, humans cannot detect pheromones because of the degeneration of the detecting organ. However, this statement seems inconsistent with the evidence that there is a chemical form of communication between women, as shown by the phenomenon of the synchronization of periods, that is, menstrual synchrony.

In 1971, Martha McClintock, at Wellesley College, published a Bachelor degree dissertation in *Nature* (McClintock, 1971). She found that female college students living together in the same dormitory have signs of menstrual synchrony. According to collected data, the menstrual cycles became 33% synchronized after 7 months. The phenomenon does not occur among women who do not share the same room. This is scientific proof that such a phenomenon exists among women who live together, such as mothers, daughters, sisters, and lesbian couples. She hypothesized that this phenomenon occurred when friends interacted with each other, as their pheromones were transferred to each other.

Twenty years after her first study of menstrual synchrony, in 1998, McClintock published, with her student, more specific experimental

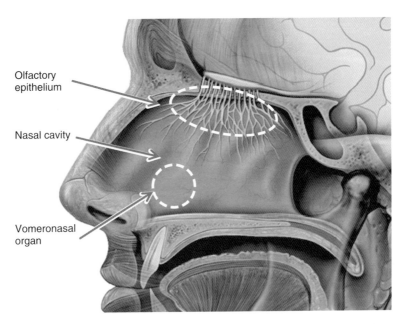

Figure 7.2   The vomeronasal organ as a remaining organ in a nasal cavity. Modified from Wikipedia Commons (http://commons.wikimedia.org/wiki /File:Head_olfactory_nerve.jpg) (accessed 17 July 2014).

In the figure, labels point to: Olfactory epithelium, Nasal cavity, and Vomeronasal organ.

data proving that human pheromones exist (Stern and McClintock, 1998). They showed that there are odorless pheromones which affect the ovulation period. The research team conducted the experiment on 29 women with regular menstrual cycles ranging in age from 20 to 35. They selected nine women and asked them to wash thoroughly. Then, they put a pad under their armpits for at least 8 hours. In this way, the research team collected a specimen from each phase of the menstrual cycle. After the collected pads had been treated with alcohol to eliminate the smell, they were then held under the noses of the rest of the women. They found that with the majority of the subjects, the menstrual cycles either slowed down or lengthened. The odorless compounds from the armpits of women in the late follicular phase shortened the menstrual cycles of recipients. The underarm compounds from the same donors that were collected later in the menstrual cycle (at ovulation) lengthened the menstrual cycles. By showing, in a fully controlled experiment, that the timing of ovulation can be manipulated, her study provided definitive evidence of human pheromones.

Human sexual behavior is not normally affected by olfactory stimulation. On the other hand, many animals are sexually motivated by smell. Androstenone is similar to the smell of musk with a little bit of urine. Alpha-androstenol, a similar substance, is a steroid that smells like musk and is found in the testicles or the saliva of boars. This stimulates the wild female swine. Truffles, which are edible fungi, also contain this substance. Thus, a female pig can smell a truffle and identify its location even from far away.

People have many sebaceous glands all over their bodies. These glands produce many substances that gather around hairs and are released with sweat in the form of oil. This oily fluid is composed of 12 types of odorous steroids, of which androstenone is the largest portion. However, since not everyone can smell this substance, it is very unlikely that it can be used as a chemical to initiate sexual attraction between people. Some women can sense a slight smell when it is in a concentrated form. There is even an anecdote that shows that women unconsciously respond to this substance.

In the late 1970s, Tom Clark of Guy's Hospital in London conducted an experiment by spraying androstenone onto the seats at a theater, which were not reserved, and compared the response of the audience to seats that were not sprayed. Even here, women preferred the seats where androstenone had been sprayed. Clark (1978) wrote an article in *World Medicine* about his pheromone experiment at the theater.

"We booked the Gulbenkian Studio Theatre in Newcastle for the happening. We studied the audience which came into the afternoon and evening shows and got a good idea of the seating patterns. The Gulbenkian studio, part of the university theatre complex, is used for informal experimental theatre. Most importantly, all the seats are unreserved. We decided to spray certain seats and to have some sort of stage experiment. Before the evening performance we sprayed the stairs to one of the balconies and the chairs on it with androstenone. After the audience had come in, the chairs were found to be occupied mainly by women. It had worked much better than we dared hope."

Androstenone can be detected even at 1 part per billion (ppb). Women can smell it ten times better than men. Dr George Preti of Monell Chemical Senses Center conducted a pheromone experiment using a male armpit extract (Preti *et al.*, 2003). The substances collected from men's armpit sweat, presumed to be pheromones, were given to women who were asked to smell the test sample for 6 hours. They did not recognize that it was men's sweat and were more relaxed and less tense after than before the experiment. Women who are relaxed are more likely to form a relationship with a man and to change the length and timing of the menstrual cycle, which affects fertility.

If there is one thing that can shake up even the most bold and cool-headed man, it's probably a woman's tears. There is even an expression that a lady's tears are her best weapon. A recent research experiment showed that the reason why a man becomes weaker in front of a woman's tears is because of the chemical reactions in a tear. Dr Shani Gelstein's research team made 50 men, ranging in age from 24 to 32, smell women's tears (Gelstein *et al.*, 2011). After sniffing the tears, men experienced reduced levels of testosterone. The team conducted a comparison test with saline solution that had the same salt concentration as the tears from a woman who had watched a sad movie. After the participants sniffed the tears and saline, their physiological arousal, such as galvanic skin response, heart rate, and skin temperature, as well as their internal testosterone levels, were measured. The test results showed that testosterone concentration level in the blood was decreased by 30% after sniffing tears. Galvanic skin response and heart rate were also visibly decreased. This meant that men reduced their aggressiveness and calmed down. Testosterone is a representative male hormone and is closely related to aggressive behavior. On the other hand, the concentration level of testosterone was not changed after sniffing saline.

These phenomena seem inconsistent with the statement that people cannot detect pheromones due to the deterioration of the human vomeronasal organ. However, this contradiction makes sense if people sense pheromones though a different organ, other than the vomeronasal organ. For example, a newborn rabbit finds its mother's breast by responding to pheromones. A female pig responds to a male pig's pheromones through its olfactory organs, not the vomeronasal organ. It would be shocking if humans were the only species that had lost this ability to smell. In 2001, Professor Peter Mombaerts and his research team at Rockefeller University found pheromone receptor genes in mammals (Del Punta *et al.*, 2002). They revealed that the pheromone receptor genes are developed in the olfactory membrane. This affirms what has long been just a theory: that humans detect pheromones through the olfactory system not through the vomeronasal organ. Most of the 135 pheromone genes are fossilized and only five remain. However, people normally do not react basally as a result of pheromones. However, a young couple in a closed room may bring out the hidden animal nature in man.

Another set of experiments, conducted by Haviland-Jones and colleagues (2013), showed how a non-detectable level of flower scent influenced people's behaviors. The researchers placed a pellet that would release a floral (gardenia) smell or a pleasant manmade perfume in a controlled manner. Participants were not able to detect the peri-threshold odors. As a control, the third room had fresh, non-scented air. College students ($n = 59$) entered one of the three rooms and wrote about life events: one from the distant past, one from the recent past, and one future event. The essays were coded for positive emotion words. In the second part of the study, the participants entered another room containing a mime and were asked to direct her to act out an emotion from the event in the distant past. Participants in the floral room used about twice as many positive emotion words as those who had experienced the other two conditions in their writings about recent and future events; there were no differences for distant past events. Most of those from the floral room (74%) were likely to approach or touch the mime compared with 20% from the fresh air room and 40% from the perfume room. Dr Haviland-Jones (2013) at Rutgers University stated, "The presence of a subtle flower scent affected the students' thoughts and actions."

Smells also play big roles in a first impression. Dr Nicole Hovis and Theresa White, of Le Moyne College in Syracuse, NY, conducted one experiment with 65 female college students (Welsh, 2011). They showed the shadow of a person of an unknown gender and made the students smell a small bottle (onion, lemon, or water). Then, the students were asked to guess the gender and fill out a survey on the personality of the person in the shadow. The result showed that students who smelled onions saw the shadow as belonging to a man, while students who smelled lemon judged the person of the shadow as a clean and neat woman. This proves that smells can influence a first impression.

The Healing power of aromas

The widely acknowledged effect of smells generally comes from the influence of reinforcement. For example, there is a negative placebo effect for making new building syndrome worse. This syndrome is related to how much tenants believe that the stale smell in an office is from harmful building materials. On the other hand, there is also a positive placebo effect such as the popular aromatherapy. Aroma therapists say that aroma positively changes feeling. For example, lavender oil relaxes tensions, while neroli oil (blossom oil of *Citrus aurantium*) is stimulating.

However, recent research proved that false information can reverse the therapeutic effect of these two fragrant oils (Estelle Campenni *et al.*, 2004). If the test subject smelled lavender and was told that lavender has a relaxation effect, the person actually became calm and the heart rate and skin conductance level were changed. On the contrary, once the person was told that lavender has a stimulating effect, the person quickly became very excited. There was the same reversed result with neroli oil. Simply dropping a good word leads to a measurable positive placebo effect in aromatherapy.

If we smell an odor ingrained in our memory from a happy moment in our childhood, we can easily go back to that joyous feeling. Using these odors can help human relationships and social activities by bringing emotional stability. There are few treatment methods that have a general effectiveness as strong as the placebo effect. Aromatherapy wakes up the therapeutic power within us when a fragrant substance triggers our own healing ability. Just smelling coffee also has some healing effects. People who do not like caffeine or the taste of coffee can just inhale the smell of coffee for their brain to become more stimulated and stress-relieved.

Aromatherapy

Aromatherapy is a treatment method that uses odorants of herbs and other natural plants to maintain the health of the mind and body. Normally, aromatherapy uses essential oils extracted from plants, but it also broadly includes the utilization of other plant odorants that bring positive results to the mind and body, such as drinking herbal tea or smelling flowers. Humans actually have an innate healing power. There are some treatment examples that enhance a weakened healing power with aromas, which may include situations where the feeling of fatigue is reduced after smelling a smooth and sweet aroma or feeling better when one drinks refreshing tea.

The word aromatherapy was first introduced by the French chemist Gattefosse in his book *Aromatherapy* in 1928; however, the utilization of odorants in plants in daily life began a long time ago. In ancient times, fragrant odors were thought to be given by the gods, and thus junipers or flowers were burned in religious services. This goes back to 3000 BC. In ancient Egypt, essential oils were already being used for medicinal purposes or as cosmetics, while cedar, myrrh, and cinnamon were used as preservatives in preparing mummies. According to papyrus documents, odorous plants such as frankincense (also called olibanum), oregano, and coriander were being used as incense or perfumes. Not only in Egypt, but also in Mesopotamia and Greece, people used odorous plants in religious services or for medicinal purposes. In India, Ayurveda (traditional medicine) was established around 600 BC and continues today.

Around the tenth century, Arabians invented steam distillation, which is still used today. In the twelfth century, lavender was harvested in England and lavender water, which is distilled incense water, became very popular. Starting from the twelfth century, odorous plants were submerged in oil and the oil was then heated to extract the odorants for practical use. Rosemary alcohol water (the origin of eau de toilette) was considered as water for juvenescence by Queen Elizabeth of Hungary. She used it to wash her face.

In 1664, in London, the black plague was spreading and the world came to learn about the disinfectant effects of odorants. In eighteenth century France, the fragrance industry grew around Grasse, which is the town known as the world's perfume capital. In Europe, medical treatments using essential oils or odorous plants became popular. Various aromatherapy treatments were declined in the nineteenth century as the result of advanced Western medicine and pharmacy; however, recently, they have become popular again all over the world.

Aromachology

Recently, many different types of fragrant ingredients have been used in cosmetics and in bath products. However, since the word aromatherapy infers the association with medical or pharmaceutical treatments, there were opinions that the terminology was not fit for cosmetics. Thus, instead of the word aromatherapy, aromachology has been used, which is a general term to encompass the scientific area that explains the psychological effects of fragrances. Aromachology is the science that explores the mutual relationship between fragrance technology and psychology, which stimulates the olfactory circuit in the limbic area of the brain and arouses various emotions and feelings such as calmness, hyperthymia, sensitivity, happiness, and peacefulness. Because it is possible to record every measurable step, there are very accurate statistical analyses available for aromachology. Therefore, just proving individual or subjective cases for the effectiveness of fragrant materials on brain functions is not a part of aromachology. It does not deal with the physiological effects other than psychological functions of active biochemicals in the blood streams, which are introduced by dermatologically absorbed fragrances through massage or inhalation. Also, aromachology involves various volatile compounds and is not limited to natural oils but covers a wide range of volatiles, including compound flavors.

Because the body and the mind are essentially related, emotional or psychological problems greatly affect physical health. For example, excessive stress will cause changes to the mind and body, which will break the hormone balance, decrease metabolism, and lead to skin trouble. Since the limbic system is the center for controlling emotions, it is very sensitive to smell. Therefore, it is possible to manipulate a person's emotions and actions with aromas. An aroma that gives a positive image can change a person's behavior positively and provide joy. For example, psychologists found out that regularly injecting mint or lily flavor into the study room of college students increases attentiveness and learning performance. Vanilla flavor was revealed to relax patients who are taking MRI scans. In addition, pleasant aromas raise self-esteem and increase work performance in workplaces. Typically, lemon flavors are known to help reduce mistakes during data input or word processing.

In general, there is a correlation between particular aromas with specific body functions. For example, while lavender flavor calms people down, basil flavor stimulates. However, subjective psychological factors can affect physical responses more than aromas do. It is very difficult to accurately determine how each individual will respond to a particular aroma. This is because the effect of the given aroma can be related to a particular emotion or psychological preference of an individual. The aroma that is related to an individual's emotions, memories, and psychological status affect the results more significantly and increase the deviation in the test results. A normally unpleasant odor might bring positive results to some people. For example, the odor of geranium can calm down nervous people in some cases, while it can make others more anxious. A brainwave pattern detector is a useful instrument in aromachology studies. In a relaxed state, the alpha wave increases, while the beta wave increases in a stimulated state. In 1979, John Steele used a brainwave detector to measure the influence of essential oils on brainwaves (Horowitz, 1999). Through this research, lavender and sandalwood were found to have a calming effect, basil had a stimulating effect, and rosewood and geranium had a balancing effect. In addition, an electroencephalograph (EEG) depends to a great extent on the expectations and experiences of the examinee with the specific aroma.

Phytoncide

Across the ages, people have thought that forests are a good place for relaxation. Up until the beginning of the twentieth century, the only treatment for tuberculosis, which took many lives, was to recuperate in a forest. Many patients actually experienced great success. The benefit of phytoncide is that it has a wider spectrum for pathogens and more practical applications than regular antibiotics, and that it is easily absorbed into the body without any side effects because it is a substance readily available in natural forests. Pathogenic microorganisms cannot survive in a forest full of phytoncide. Therefore, if a patient recuperates in a forest, there is less chance of a secondary infection from other pathogens. The places with the greatest amounts of pathogens are hospitals and other places where there are a lot of people living together.

Phytoncide is a general term for antimicrobial substances secreted by plants. It does not refer to just one substance. It includes terpenes, phenolic compounds, alkaloids, and glucosides. Every plant has antimicrobial substances and, thus, possesses phytoncide in different forms. Generally, the antimicrobial substance from a healthy higher plant (i.e., vascular plants) is called phytoncide, and the strong antimicrobial substance secreted from infected plants after invasion of disease microorganisms is classified as phytoalexin. The antimicrobial phytoncide is not a strong microcidal substance that kills the disease microorganisms in a short time, but it is mainly a suppressive microstatic substance at a preventative level. Thus, only a long-term green shower in a forest can be effective for skin trouble, asthma, and tuberculosis. In most cases, people take a green shower in a forest to strengthen their mind and body, to prevent a disease, and maintain their healthy life.

Among the chemicals we come into contact with during a green shower, there is an important substance called terpene. It is a pungent aromatic substance that includes the well-known alpha-pinene, amongst tens of others. Although in general phytoncide is an antimicrobial substance used against disease-causing microorganisms, terpenes have other multiple functions, which include self-activation of some biochemical reactions in plants, attracting agents for various insects, and inhibition of the growth of other competing plants. When terpenes are absorbed by the human body, they increase metabolic reactions by stimulating the skin, helping blood circulation, relaxing psychological status, and acting as antimicrobial agents. Therefore, people achieve the positive therapeutic effects of a green shower through the various active functions of terpenes as well as the activity of phytoncide. Forests also provide great satisfaction to all of our senses including vision, smell, taste, hearing, and touch; therefore, a green shower is a very effective method for strengthening the condition of our emotional and psychological health. If you compare the rate of recovery from fatigue with and without alpha-pinene, there is a significant difference in the recovery rate. The rate of recovery is faster and there are less signs of fatigue the next day when sleeping with alpha-pinene.

Is geosmin foul or pleasant?

Geosmin, originating from the Greek word for "earth odor," is a simple molecule that causes the characteristic smell of soil. The threshold value in water is about 0.0082–0.018 ppb, which is very low, making it difficult for it to be to totally eliminated. This earthy odor is actually very familiar to us. This is the smell from dried dirt roads after a sudden rainstorm. Also, it is the earthy odor when we rototiller a a dry filed or pull up weeds. However, the soil itself, which is mainly composed of silicate, does not have an odor. Red clay soil contains high levels of oxidized iron, although it is hard to recognize the metallic smell from the mud because the smell of oxidized iron is so faint. Almost all of the smell from the soil is actually due to the odor of geosmin, which originates from bacteria such as *Streptomyces*. As a member of the Actinomycetes family, it is found in the ground and produces mycelium and spores, such as molds. Humans are very sensitive to geosmin. The earthy odor can be recognized as soon as the concentration of geosmin in the air goes above 5 parts per trillion (ppt). Precise analytical instruments are unable to detect concentrations as low as 5 ppt, so the human nose performs better than expensive laboratory equipment.

Geosmin is not harmful to the human body. However, nobody likes the smell of soil in their tap water. Water treatment facilities that produce tap water try to maintain the geosmin concentration at less than 20 ppt so as not to disturb the residents. Nevertheless, it is not easy to eliminate such a dilute trace amount of geosmin at the water treatment plant. This is actually carried out more easily in the home. Geosmin is volatilized when it is heated; therefore, simple heating of tap water will eliminate the earthy odor. However, some people enjoy this smell. People who spent their childhood in the countryside or in a farming area recollect the days of their childhood when they smell this earthy odor. Recreational forests in a southern province of South Korea, Jeollanam-do, have become areas for the mass production of geosmin, which is helpful for mental strength and depression. In August of 2012, Jeollanam-do Institute of Health and Environment analyzed the soil samples of six recreational forests (Kim *et al.*, 2013). Geosmin levels were shown to be higher in the soil of cypress forests (136.1 µg kg^{-1} from Chugryeong Mountain, 62.0 µg kg^{-1} from Paryeong Mountain, and 38.5 µg kg^{-1} from a Woodland Recreational Forest) than in pine forest soil (41.2 µg kg^{-1} from Baega Mountain, 15.7 µg kg^{-1} from Baekgun Mountain, and 7.2 µg kg^{-1} from Juknokwon). When the researchers analyzed the brainwaves of those who entered the forests before and after inhaling geosmin in order to reaffirm its emotional stabilizing effect, they saw an increase in alpha and delta waves after inhalation of geosmin, indicating a stabilized mind and body. Furthermore, they observed increased brain relaxation, activation, and concentration with a decreased stress index. Geosmin, in addition to phytoncide, has become a substance that explains the natural healing effects of a green shower in recreational forests. The same substance can become a foul odor if it is contained in water but a pleasant aroma if it is in a natural forest.

As an example, how do camels find oases in the desert? Dr Keith Chater suggested a theory that camels use their olfactory sense (Royal Society of Chemistry, 2012; Simons, 2003). He has suggested that camels can detect the odor of geosmin that has been released by *Streptomyces* miles away in wet ground and they track the geosmin odor to find an oasis. The camel could also carry away the spores of the *Streptomyces* bacterium and disperse them at the next water hole.

Multiple chemical sensitivity (mcs): there are people who are really intolerant to odorous chemicals

There are some people who suffer hypersensitive reactions to various, or to all, odorous substances. Some of them are so sensitive to chemical substances that their symptoms show up with the faintest levels of perfumes. Their pain is so severe that sometimes they give up going out and stay at home to avoid coming into contacting with anyone who uses perfume and other fragrance products in general. One woman moved to the Arizona desert with her family and lived inside a custom-designed trailer house made of nontoxic metals because she hoped that an isolated life could solve her problem with chemical odorants. MCS patients have various symptoms including breathing difficulty, pains in the throat, chest, or abdominal region, asthma, skin irritation, contact dermatitis, hives or other forms of skin rash, headaches, neurological symptoms, tendonitis, seizures, visual disturbances, extreme anxiety, panic and/or anger, sleep disturbance, suppression of immune system, digestive difficulties, nausea, indigestion/heartburn, vomiting, diarrhea, joint pains, vertigo/dizziness, an abnormally acute sense of smell, sensitivity to natural plant fragrance or natural pine terpenes, insomnia, dry mouth, dry eyes, and an overactive bladder. The physiological causes of multiple chemical hypersensitivity and odor hypersensitivity have not been identified so it is also called a non-specific environmental hypersensitivity. Even though there is no apparent cause, there is one commonality that is found for all of the patients. They all claim that they are more sensitive to smell than other people. However, when specifically controlled scientific experiments were conducted to identify their odor sensitivity compared with normal people, no significant difference was found. It was simply concluded that their psychological response of how to perceive the odor was different from ordinary people. MCS patients perceive the rose smell as less pleasant than normal people and say their eyes, nose, and throat hurt. In a different experiment, MCS patients detected odors at considerably lower levels than ordinary people, but they also responded to the clean air used as a placebo. This means that the problem is not their physiological sensitivity to chemical substances but their psychological hypersensitivity concerning danger and anxiousness.

Warnings and campaigns about environmental pollution are important and helpful for protecting the environment, but in some cases they raise unintended sensitivity to harmless environmental substances. Odor hypersensitivity to our surroundings is not common; it has been known for 100 years. Eugene Rimmel, the perfumer in London who first invented mascara, said that there are many people who imagine that perfumes are harmful (Rimmel, 1867). One woman, who argued that she could not bear any smells, fainted when her friend brought her a rose. However, it turned out that the flower was not real. The fateful rose was artificial. Despite such incidences, some MCS patients deny this psychogenic hypothesis. They believe that the problem is caused by the chemical substances and become angry when they hear that the problem lies in their psychological status, because accepting the psychogenic hypothesis means that their pain is imaginary.

At one time, the Chinese restaurant syndrome was widespread due to the side effects of MSG (monosodium glutamate). Even though multiple scientific experiments clearly revealed that it had nothing to do with MSG and the US Food and Drug Administration (FDA) announced that MSG was harmless, some people are still reluctant to go to Chinese restaurants or to eat food products that contain MSG. MSG is dissociated into glutamic acid and sodium ions. Glutamic acid accounts for the highest percentage of amino acids, which are the building blocks of proteins. Umami is a taste perceived through the detection of glutamic acid by the taste buds on the tongue. There is no technology in the world that can distinguish the origin of the glutamic acid, as to whether it comes from hydrolyzed proteins or from MSG. Indeed, hydrolyzed proteins also contain various other amino acids in addition to glutamic acid, so their glutamic acid taste might be different from MSG. However, when someone feels that his/her body is over reacting to MSG, it is worth considering whether the person is thinking about MSG too much. Our bodies can easily become hypersensitive if we wish them to be.

There are many people who die from anorexia after becoming psychologically sensitive to foods. Anorexic patients become overly afraid of becoming obese and are tortured by their imaginations, where the food they consume is sticking to their body cells. With this type of psychological condition, eating becomes very stressful. Despite their efforts to revert their minds back to normal again, swallowing food becomes too painful. Stomach cramps occur when they even eat just a small amount. Also, it is healthy to eat vegetables, but strict vegetarians sometimes find it very difficult to consume meat again. Fruits and vegetables might taste good and smell fresh to vegetarians, but meats become mushy, limp, and nasty. Even with a determined change of mind to eat some meats or even sushi, it not easy to enjoy them again.

So, camels can find oases through the smell of geosmin and enjoy the earthy smell of geosmin in forests, but consider it as an undesirable odor in tap water. However, whether geosmin is a pleasant scent or a foul odor is not up to the geosmin: it is only up to our minds.

References

Campenni, C.E., Crawley, E.J., Meier, M.E. (2004) Role of suggestion in odor-induced mood change. *Psychol. Rep.* **94**(3C): 1127–1136.

Chu, S., Downes, J.J. (2000) Odour-evoked autobiographical memories: Psychological investigations of Proustian phenomena. *Chem. Senses.* **25**(1): 111–116.

Clark, T. (1978) Whose pheromone are you? *World Medicine* July 26, 1978. digilander.libero.it/linguaggiodelcorpo/whoseph/ (accessed 17 July 2014).

Del Punta, K., Leinders-Zufall, T., Rodriguez, I., Jukam, D., Wysocki, C.J., Ogawa, S., Zufall, F., Mombaerts, P. (2002) Deficient pheromone responses in mice lacking a cluster of vomeronasal receptor genes. *Nature.* **419**, 70–74.

Gattefosse, R.M. (1937) *Aromathéraphie: Les Huiles Essentielles, Hormones Végétables.* Librairie des sciences Girardot.

Gelstein, S., Yeshurun, Y., Rozenkrantz, L., Shushan, S., Frumin, I., Roth, Y., Sobel, N. (2011) Human tears contain a chemosignal. *Science.* **331**(6014): 226–230.

Hardy, C., Rosedale, M., Messinger, J.W., Kleinhaus, K., Aujero, N., Silva, H., Goetz, R.R., Goetz, D.,. Harkavy-Friedman, J., Malaspina, D. (2012) Olfactory acuity is associated with mood and function in a pilot study of stable bipolar disorder patients. *Bipolar Disord.* **14**(1): 109–117.

Haviland-Jones, J., Hudson, J.A., Wilson, P., Freyberg, R., McGuire, T. (2013) The emotional air in your space: Scrubbed, wild or cultivated? *Emotion, Space Soc.* **6**: 91–99.

Horowitz, S. (1999) Modern applications of essential oils. *Altern. Complementary Ther.* **5**(4): 199–203.

Kim, G.S., Lee, S.Y., Ha, H., Jeong, S.H. (2013) The geosmin components of soil road well-known in Jeollanam-do. *The Report of Health and Environment Research. Jeollanam-do Institute of Health and Environment*, volume **23**, pp. 187–206.

Li, X., Li, W., Wang, H., Bayley, D.L., Cao, J., Reed, D.R., Bachmanov, A.A., Huang, L., Legrand-Defretin, V., Beauchamp, G.K., Brand, J.G. 2006. Cats lack a sweet taste receptor. *J. Nutr.* **136** (7): 1932S–1934S.

McClintock, M.K. (1971) Menstrual synchrony and suppression. *Nature.* **229**: 244–245.

Negoias, S., Croy, I., Gerber, J., Puschmann, S., Petrowski, K., Joraschky, P., Hummel, T. (2010) Reduced olfactory bulb volume and olfactory sensitivity in patients with acute major depression. *Neuroscience.* **169**(1): 415–421.

Nei, M., Niimura, Y., Nozawa, M. (2008) The evolution of animal chemosensory receptor gene repertoires: Roles of chance and necessity. *Nat. Rev. Genet.* **9**(December): 951.

Preti, G., Wysocki, C.J., Barnhart, K.T., Sondheimer, S.J., Leyden, J.L. 2003 Male axillary extracts contain pheromones that affect pulsatile secretion of luteinizing hormone and mood in women recipients. *Biol. Reprod.* **8**(6): 2107.

Rimmel, E. (1867) *The Book of Perfumes*. Chapman and Hall, London, pp. 13–14.

Royal Society of Chemistry. (2012) Geosmin. Chemistry in its Element: Compounds, Series 2: *Chemistry World Podcast*, **14** December 2012.

Simons, P. (2003) Camels act on a hump. *The Guardian.* March 6, 2003. http://www.theguardian.com/science/2003/mar/06/science .research/print (accessed 17 July 2014).

Stern, K., McClintock, M.K. (1998) Regulation of ovulation by human pheromones. *Nature.* **392**: 177–179.

Welsh, J. (2011) Smell of Success: Scents Affect Thoughts, Behaviors. *LifeScience.* http://www.livescience.com/14635-impression-smell-thoughts-behavior-flowers.html (accessed 17 July 2014).

8 Taste is Regulated by Flavor, and Flavor is Regulated by the Brain

The sense of smell is directly connected to the imbic system, in other words, to survival and emotion

The olfactory sense, unlike vision or hearing, is directly linked to the limbic system. The amygdala is the center of emotion. Stimulating the amygdala with an electrode can cause almost every psychological condition that a person can perceive, such as intense anger, affectionate love, and disheartening sadness, almost instantaneously. Emotion is an important mechanism for motivation and desire. If the hippocampus, which has a central role in memory, is damaged, people cannot remember new smells (i.e., information). The memory prior to damage to the hippocampus is completely preserved, but new memories cannot be saved.

Information coming from the corpus striate goes to the thalamus, is relayed to the cortex, and returns to the striate, creating a large closed circuit of neuro-signal transduction. The frontal lobe receives information relating to smells directly from the cortex, along with related information on the body's response and this information can affect the next action. Memories of a particular smell or action and their related experiences help people learn possible reactions to various actions. The memory of the previous experience is used when selecting the next action. The synaptic connection between the cerebral cortex and striate is continuously being modified through experiences of success or failure after seeking or avoiding specific food, animals, and environmental odors. Thus, if you are faced with a similar situation, the striatal–thalamic–cortical circuit, formed by

How Flavor Works: The Science of Taste and Aroma, First Edition.
Nak-Eon Choi and Jung H. Han.
© 2015 John Wiley & Sons, Ltd. Published 2015 by John Wiley & Sons, Ltd.

hundreds or thousands of experiences, remembers the last action that was made after the previous encounter with the odor. Because a cyclic circuit is made with this structure, the neuro-signals circulate through the vast frontal striatal–thalamic–cortical circuit for a relatively long time (a couple of seconds), resulting in an amplified action and the current circuit is then stored as a memory. Figure 8.1 summarizes the mechanism of sensorial perception in the brain.

Food is composed of a variety of individual flavors. If a person smells the odor of a dish, the neurons in multiple areas of the brain light up. This shows that these mixed odors are simultaneously activating different olfactory receptors. We can smell roasted meat, flour fried in butter, sweet vegetables, tomato paste, and spices. However, how do all of these components converge to a specific, representative taste of one dish? This is known as the integration of information from hundreds of olfactory receptors into one perceived smell. Let's say that a tender filet mignon with a sauce and mashed potatoes full of butter on the side are on a table. Many aromas, such as the brown smell of grilled meat and the unique starch smell of mashed potatoes, bombard our noses. From these foods, we can take in the aroma of the whole dish or each individual odor. In other words, we can experience

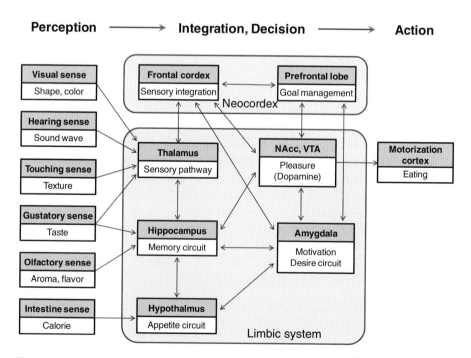

Figure 8.1   Mechanism of sensorial perception related to foods in the brain.

a food symphony of a multispectral aroma that overlaps all odors or analyze the data from the sensory input to focus on the odor of the potatoes, the sauce, or roasted beef individually. As such, selective focusing on a specific perception is carried out without any effort or training, but how this occurs in our brain is still unknown. Problems with the separation and the combination of multiple perceptions cannot be solved by simply tracing the route the signal transduction takes in the nerves in the brain. Despite the precise mapping of human brain activities by brain scientists, it is difficult to explain how we can integrate all perceptions or how we can freely go back and forth between different individual sensory perceptions, for example, distinguishing the feeling of food as a whole or the individual smell of the potatoes, the sauce, or the beef.

Many functions that are necessary for brain activity are created during the developmental stage of the olfactory functions, such as controlling the magnitude of the signal by feeding back the signal from the olfactory cortex to the olfactory bulb (i.e., adaptation or olfactory fatigue), amplifying or diminishing the loop signal according to the selected odor and eventually strengthening the synaptic connection to remember and learn (i.e., control of emotion and memory), comparing and stimulating (i.e., new input) the signal pattern with the saved (i.e., remembered) pattern, amplifying the selected or desired signals while reducing noise, conjecturing the whole from limited information by filling in the information gaps, and connecting the olfactory sense to other senses. Perceiving smell uses the same mechanism as when recognizing a face through vision, and the olfactory circuit is similar to the hippocampal circuit. It is a typical characteristic of evolutionary development that a successful method is used repeatedly and is modified little by little. In order to understand ourselves more, we must make better efforts to understand our brains.

Neuroplasticity in the brain

Neuroplasticity is a concept where the functional area of the brain is modifiable, similar to clay or plastic. A long time ago, it was believed that brain cells last for a lifetime and conduct only certain designated functions, so if the brain is damaged, the damaged part cannot be restored. With the discovery of the Broca's area and the Wernicke's area in the brain (which are in charge of language), it became clear that each part of the brain plays a distinct role. After further scientific advances, it was realized that the modification of assigned functions was possible to a certain extent. People with permanent damage to the Broca's area could still talk and animals with their motor cortex surgically removed could still move. The visual cortex of a blind person can even be restructured to process audio signals. The volume of the brain of an animal that grows in an environment with plenty of stimuli is bigger and it has been observed that new brain cells are produced even in old age. Thus there is now clear evidence for neuroplasticity in the brain.

Neuroplasticity can sometimes cause phantom limb syndrome. A phantom limb is the sense that an amputated or missing limb is still attached to the body and can move like other body parts. Most of these senses are painful and it is frustrating to have a non-existent hand that is itchy or is hurting, but there is nothing that can be done about it. However, they say that scratching another particular body part, such as part of the face, makes it feel like the itchy part of the phantom limb is being scratched. This is because the nerve cells that were responsible for the sense in the hand have moved away to the nerves in the face as there were no more stimulating signals.

Even though neuroplasticity leads to such phantom pain and phantom limbs, it also plays a very useful role. If necessary, it can allow people to "see" sound or "hear" light. A British soldier, who lost his sight in the war in Iraq, in a rocket-propelled grenade attack, partially recovered his sight through his "tongue" with the help of state-of-the-art technology. The difficult life of Private Craig Lundberg changed when he was selected to be the first British experimental participant in the United States' cutting edge technology known as BrainPort (Figure 8.2). BrainPort is a piece of equipment that allows the blind user to detect an object that is right in front of them. This equipment connects camera sensors from a pair of sunglasses to a device that delivers electrical signals to the tongue. The principle is that after the camera has detected the objects in front of it, the electrical signals of the camera images are sent through the tongue device. This electrical signal forms a black and white visual signal inside the brain through the stimulation on the tongue, without going through the optic nervous system. It is not yet perfect, but not only can the user distinguish the shape of the objects, they can also read simple characters.

Is synesthesia a malfunction or a blessing?

The phenomenon of simultaneous detection of more than one of the human's five senses is called synesthesia. This is caused by a

Figure 8.2   **The BrainPort V100 includes a video camera mounted on a pair of sunglasses. The camera works in a variety of lighting conditions and has an adjustable field of view (zoom). The tongue array contains 400 electrodes and is connected to the glasses via a flexible cable. Black pixels from the camera are felt on the tongue as strong stimulations, there is no stimulation of white pixels, and gray levels have medium levels of stimulation. Users report the sense as pictures that are painted on the tongue with tiny bubbles. Image courtesy of Robert Beckman, CEO, Wicab Inc., Middleton, WI, USA.**

crossover of adjacent independent sensory areas of the brain. These sensory regions have several connections over many areas, making cross-wiring distinctly possible. Synesthesia happens when an appropriate clustering in the nervous system of the fetus does not occur during the development of connections between the areas of the brain, which can happen in 1 out of 2000 people. People with synesthesia might see colors when they are listening to music or feel tastes when they are looking at colors. It prevents each sense from functioning accurately and individually. However, from an artist's perspective, this phenomenon can be used as a motivation for creativity. For example, colored hearing is a phenomenon of feeling a color from sounds. According to the Russian Press (Yastrebtsev, 1908), Russian composer Nikolai Rimsky-Korsakov perceived a typical colored hearing phenomenon from musical chords, such as, C major made him see a white color, and for B major he envisaged a gloomy dark blue. Furthermore, the perceived color changed when the musical key changed to a minor from a major.

There is a color–letter synesthesia as well. This refers to a situation where a person feels color from the alphabet or numbers. For example, a person with this synesthesia can see the color red from a number five written in black ink on a white background. This is because the number cognition area and color perception area are activated at the same time. For a person without synesthesia, only the relevant brain areas become activated when the person is looking at a letter or a number. However, for a person with synesthesia, the color perception area is also activated. A person with synesthesia also shows a higher performance when sensing color than a normal person, with much more activation in the color perception area of the brain. Synesthesia also moderately affects comparative and creative abilities. It is no coincidence that the frequency of synesthesia among artists is eight times higher than in groups of normal people.

Taste is a typical phenomenon of synesthesia and neuroplasticity

Synesthesia and neuroplasticity are seen as extraordinary phenomena of the brain, but taste is actually a typical example of both. The sense of taste is the integration of olfactory, gustatory, hearing, and visual senses combined in the orbitofrontal cortex, and as such is the most multi-sensory phenomenon among the brain's functions. The brain judges a taste from the effects all of the sense both directly and indirectly. In the orbitofrontal cortex, there are many nerve cells that are stimulated by taste and smell at the same time. Taste is even linked with the somatosensory system, in order be able to cognize the higher viscosity of a sweet liquid and the lower viscosity for a sour liquid. Therefore, it is actually a form of synesthesia when we produce saliva and feel sourness after hearing a story about lemons.

So taste perception is the result of synesthesia and neuroplasticity. The lifespan of a taste cell is merely a couple of weeks while for an olfactory cell it is two months. The sense changes as new cells are born every day. There is no area that has more neuroplasticity in the brain than the olfactory bulb where all olfactory signals join together for the first time just prior to transmitting to the brain. A neuron becomes more sensitive the more it is stimulated. The survival of a cell depends on its activity. A cell that is stimulated and delivers sensory signals utilizes more energy and becomes more active, while others wither away. If there is no consistent stimulation, even a sensitive area returns to its previous state. The professional ability of a perfumer, a flavorist, or a sommelier is possible because huge amounts of training and tasting have awakened the neuroplasticity of their brains, which exists instinctively, it is not a special gift in any way. The brain will begin to be more powerful if we simultaneously study the basic theory of something and then put the theory to use in practice.

Orbitofrontal cortex: where sight, taste, smell, and touch meet

The orbitofrontal cortex is located beneath the frontal lobe, the region above the eyes. In order to understand why taste is affected by various sensorial factors, we learn more about the orbitofrontal cortex, where several of the aforementioned loops meet. The prefrontal cortex, which makes humans have humanitarian values, is a neocortex of the limbic system. It controls reward and punishment, in other words desire and motive. It is also responsible for regulating emotional data according to the specific situation and carrying out socially appropriate actions. The outer area of the orbitofrontal cortex is activated in a situation related to punishment, while the inner area is activated in a situation related to reward. If the orbitofrontal cortex is damaged, the patient becomes focused on more immediate rewards, obsessing over high-risk rewards rather than low-risk rewards. Also, the patient becomes irresponsible and acts in a socially inappropriate way. The ability to learn from their own mistakes disappears as well.

The orbitofrontal cortex is the secondary olfactory cortex where the processed signals of taste, vision, and touch meet. Of course taste and visual signals are processed at their particular cortex differently through several steps and loops, but a portion of the nerves gather together at the orbitofrontal cortex and integrate information collected from different senses. The cells here connect olfactory information with taste signals and visual signals, so they are often stimulated simultaneously. This is called the integration of the senses or synesthesia. Figure 8.3 shows the processes from sensing through to integration of sensorial signals. Smells, however, can be classified as either as pleasant or unpleasant types. The pleasant type is processed at the inner area of the orbitofrontal cortex, while the unpleasant

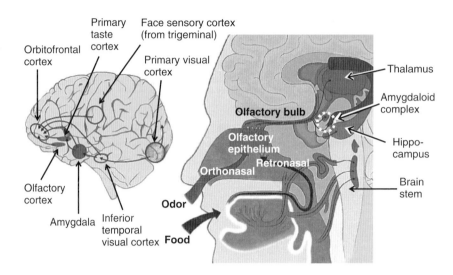

Figure 8.3 **From sensing to integration.**

type is processed at the outer area. Such classification is related to the reward and punishment system and is used as a tool for food selection. This integration, classification, and reward, which can be changed through experiences, including culture, are the motivation for developmental evolution and the reason for sense. This area also plays an important role in experiencing beauty from the arts or music, or sharing the feelings between a mother and a newborn baby.

Taste is a part of pleasure, and that pleasure becomes a part of taste

Olfactory and gustatory signals are first delivered to the amygdala and the hippocampus. These signals eventually circle the loop for amplification by the amygdala while the brain decides to remember them or not. If a positive or negative emotion is added to the signal in the amygdala, the signal becomes stronger. A strong signal is remembered through the hippocampus. Perceiving a sense is followed by a decision and a responding action (e.g., movement). If there is no decision or a responding action is not required, the act of sensing would be an unnecessary waste. Information does not have to travel all the way to the brain and back if no decisions need to be made. Neurons can receive a signal and respond without informing the brain. This process is called reflection. The brain decides and selects the appropriate signals from the received signals. It is beneficial for survival to send an appropriate signal (i.e., output for movement) after comparing the input signal with memorized information. If a brain remembered every sense, it would have storage space problems. It is advantageous to load an emotion into a meaningful signal and remember it after amplification. Emotion and pleasure were probably born through this process.

The mechanisms for the desire for food, sex, and addictive drugs are probably not completely different from one another. Consuming food for survival or partaking in sexual activity for reproduction both release signal-transmitting agents that cause pleasure. Pleasure derived from the signal transduction of the reward circuit, in other words, "reward for action," is really the driving force for developmental evolution. The reward system secretes huge amounts of dopamine. Among several types of pleasure, that to survival and reproduction is the most fundamental. The brain creates pleasure by utilizing many substances such as dopamine, serotonin, and noradrenalin. Actually, pleasure, pain, and addiction are all caused by the brain. Opioids are a group of neurochemicals that bind to the opioid receptor. One of these is morphine. The opioid receptor is a typical G-protein coupled receptor. People think it is the opioid itself that is addictive, but the chemical is not addictive. It is the brain that becomes addicted because the opioid has a similar structure to endorphin, which is responsible for pleasure. People also think that pain occurs from the wound when one gets hurt, but the wounded area only generates electrical signals. When the signals are received by the brain, the brain considers these to be harmful to survival and activates the pain circuit to create pain.

Based on most experiences in life, it is easy to see that being rational is more beneficial than being emotional. However, this cannot be true. Emotion provides true human value. If rationality simply existed without any emotion, horrible events could transpire. For example, let's say someone wanted to know the volume of blood on bleeding after a blade cut at a certain angle. This is a rational and scientific curiosity. However, what stops this experiment is the "emotion" that makes the person feel this is brutal and horrible. Emotion is also important in shopping or in marriage. Marriage is a serious life event. If someone tried to rationally calculate whether a chosen person would be the best spouse, it would be impossible to have a successful marriage. There is no evidence that shows that getting married after rationally weighing up thousands of factors is any better than getting married when swept along and making a decision based entirely on emotions. It would also be as difficult to choose the best product among the thousands for sale at the shopping mall using only rational decision-making strategies.

Experience affects taste: familiar foods are preferred

Experiences with consumer packaged goods (CPG) such as processed food products are the most critical factors to consumers when deciding whether to re-purchase products. This is why first impressions are as important as the memory of past experiences with the product. The biggest ice cream manufacturer in China is located in Inner Mongolia. This business grew and became the leading brand in the market, even compared with global brands. The brands of ice cream from Southern Chinese manufacturers also lost their market share due to the rapid growth of the Inner Mongolia manufacturer, which is located in a very cold climatic region. Mongolians have a nomadic heritage with livestock as their main livelihood. Dairy production is one of their major occupations along with meat production. One of the most critical factors for the ice cream business in the Mongolian area is that large amounts of ice cream products are consumed during the winter season. Consumers in the warmer southern areas enjoy ice cream during the summer season only, they do not consume it when the weather becomes chilly. However, in Inner Mongolia, people enjoy eating ice cream outside even at $-15\,°C$ ($0\,°F$). So although ice cream consumption in Inner Mongolia rises during summer time due to the increase in temperature, it is also popular during the winter time because the products can be sold without the need for refrigeration making it popular year round. Therefore, we can see why the ice cream manufacturer in northern China is more successful than those in southern China, thanks to seasonal consumption.

Globally, there is a bigger market for ice cream products in cold climate countries than in those with hot climates, Northern Europeans consume more ice cream than Southern Europeans. There are many different explanations as to why ice cream consumption is higher in the north than in the south but one of the most accepted explanations is the preference of the consumers and the connection to product experience and familiarity. Consumers who have a familiarity with ice want this cold ice experience during the summer time. However, consumers who have never seen ice would not look for it during a hot summer's day. Even though refrigeration has become globally popular, there are still regions with no refrigeration systems, making ice cream a rare product.

The feeling of disgust can be acquired through learning

Personal preferences vary through different experiences. All animals, including humans, have a device in the brain that rejects harmful foods. They will avoid a certain food if they have had a specific experience with it, such as a stomachache or vomiting. Taste receptors are not only dispersed in the tongue and mouth, but also in the intestinal epithelial cells. The receptors in the intestines do not feel with the same kind of taste because the signals from the intestine receptors do not go to the taste center in the brain in the same way that the signals from the receptors in the tongue and mouth do. This means that the intestine cannot feel sweetness if a piece of candy is positioned directly in the intestine with an endoscope. However, our body remembers the substance through the information integration of the intestine's signals with the sweetness signals from the taste buds. Therefore, through that same mechanism, simply imagining eating foods that caused uncomfortable incidents can upset our stomach.

The feeling of disgust plays an interesting role in the process of food preference development. Disgust is a feeling that developed to avoid spoiled and contaminated foods, especially putrefied meats. Most foods that cause a disgusting feeling, other than meats, have some common similarities. They can be foods of animal origin, such as cheese or milk, or they have very similar textures to meats. These observations support the hypothesis that the disgusting feeling originates from one of the mechanisms to avoid spoiled meat products.

Young children and toddlers do not feel disgust. They do not think their own or others' urine or feces are disgusting. Children will casually eat feces-shaped food that is made from peanut butter and smells of cheese. The feeling of disgust begins to form after the age of 3 or 4. Starting at this point, children avoid excrements and do not drink juice or milk if there is anything dirty in the cup. Once in a while, they over-react and obsessively pay attention to what is smeared on the food or the containers. We do not know why the feeling of disgust is developed. However, children, who do not learn about disgust gradually, will suddenly assume that a food is dangerous until it is proven to be safe. Children observe people around them eating meats and think that the meats those people are not eating are disgusting. Nonetheless, they readily try various new fruits or vegetables. It is probably because plants have a characteristic of not spoiling easily due to their hard cell walls compared with the cellular structure of meats. It is amazing how children intuitively know about the characteristics of cellular structure and the possibility of food spoilage with respect to the cellular structure.

Taste is affected by temperature

Georges Auguste Escoffier, a French chef, brought a revolution to the kitchen by utilizing the relationship between temperature and food in an appropriate way. In other words, he made people enjoy foods served in different courses. Everything on the menu was not prepared beforehand, but was made after it had been ordered. If the food cools down, the harmony of the various tastes is broken and the overall taste easily becomes flat. Thus, Escoffier *et al.* (1903) warn in their cookbook, *Le Guide Culinaire*, that "Customers will feel the dish is flat and tasteless if it is not served to the table warm and freshly prepared." Escoffier was referred to by the French press as "roi des cuisiniers et cuisinier des rois (king of chefs and chef of kings)".

When the food is hot, the flavoring molecules vibrate actively and vaporize into the air to fill the kitchen with a delicious smell. On the other hand, fewer volatile chemicals will vaporize from chilled foods. In this case, the taste wholly depends on the taste buds on the tongue. As anyone who has had a stuffed nose may know, the pleasure of eating is fairly dependent on smell. There are many people who presume that 90% of what we perceive as the taste of a food is actually aroma. Aroma activates the salivary gland to get ready to eat, but it also provides an indefinable complexity of taste and flavor to food. Escoffier was the first chef to utilize temperature and the sense of smell in an exceptional way.

Most bad aromas smell very strong at 26–30 °C and our nerves respond more sensitively at a relative humidity of 60–80%. Olfactory sensitivity is about 4–10 times higher in the morning than at night, with the sensitivity falling during very dry or very humid conditions, and at temperatures above 35 °C, it is impossible to distinguish mild differences in aroma. So in general, although the effect of temperature on taste is not consistent, in the main, when the temperature increases, the sweetness increases (with the exception of fructose) and the bitterness decreases. The basic taste that decides the taste of the food is saltiness and this changes the most depending on the temperature. Saltiness weakens as the temperature goes up and feels stronger when it cools down. For example, the temperatures at which food tastes the best are 45 °C for steamed rice, 70 °C for tea and soup, 80–85 °C for brewed coffee, 60–70 °C for café latte or café au lait, 65 °C for steak, 75 °C for French fries, and 74 °C for fried chicken. As a generalization, going above 70 °C makes the food too hot to eat and below 5 °C makes the food too cold to distinguish the taste. In conclusion, the temperatures that influence food tastes the best are between 60 and 70 °C for hot food and from 5 to 12 °C for cold food.

Price: expectation affects the taste

Our sense of smell is affected by the information that is processed by the brain. Parmesan cheese and vomited undigested food are both full of butanoic acid, but we react differently according to the information we are provided with: we have a good response to the smell of Parmesan cheese, while we feel displeasure at the vomit. Escoffier understood this psychological effect very well, thus he enthusiastically used deceiving power. He gave fancy names to his foods and served them on gold-plated silver dishes, having obtained a shiny clean plate collection from estate sales of aristocratic families. Escoffier never used plain names such as "steak with gravy sauce," he provided magnificent names such as "filet de boeuf Richelieu." He dressed his waiters in tuxedoes and personally directed the design of the interior of his restaurant. In short, his idea was that there must be a perfect atmosphere for a perfect dish, realizing that our senses are significantly affected by the atmosphere.

A small piece of food in our mouth is only a small part of what we think when we taste it. Equally important are all the past experiences in our brains. In 2001, Frédéric Brochet at the University of Bordeaux tested two different bottles of the same medium quality wine (Brochet, 2001). One bottle had a luxury brand name, while the other was a plain brand. When experts tasted them, they evaluated the two wines as being completely different. The one bottled as a luxury brand received the comments "Tastes good with a good oak fragrance, several indefinable and complex tastes are harmoniously balanced, and it swallows smoothly." On the other hand, the one bottled as a plain brand received the comments "The fragrance is weak and evaporates quickly. It has low proof, is flat, and tastes stale." Although we taste all flavors simultaneously, and the trained wine experts made a lousy mistake, it is not solely their fault. From the beginning, prejudices direct our brains not to distinguish actual sense from perception. If we think that a wine is cheap, we really taste cheap wine.

Prejudices are effective at distorting perceived senses

Pamela Dalton, a Psychologist at Monell Chemical Senses Center, proved that expectations actually change the perception of smell (Dalton, 2012). The participants were divided into three groups, who sat down in a lab, and were then exposed to an odor that was neither pleasant nor unpleasant for 20 minutes. To one group, she did not say anything about the odor, while she told the second and the third groups that the odor was an industrial chemical that might be harmful and a distilled pure natural extract, respectively. The result was that the subjects who were either told nothing or positive information felt the odor to be weaker as time passed. In contrast, the subjects who were told negative information felt the odor became stronger as time passed. In other words, an odor that is thought to be good disappears from consciousness fast, while an odor thought to be harmful keeps our attention and remains strong. This experiment shows that prejudices are very effective at distorting the senses.

In 1988, Edwin E. Slosson, a chemist at the University of Wyoming, pulled a prank experiment during his class in a lecture room. He explained to his students that he wanted to prove the spread of an odor through the atmosphere. He poured a bottle of distilled water over some cotton wrapping material and informed his students that a "strong and peculiar" but "not too disagreeable" odor would diffuse in an orderly manner from the front to the back of the hall. Using a stopwatch, he found that within 15 seconds, many students in the front row raised their hands in sudden recognition of a smell, and within 40 seconds, a show of hands indicated that this travelling wave of odor had reached the furthest row of seats (Slosson, 1988). The experiment had to be terminated prematurely after one minute, because of increasing somatic and visceral complaints from those in the front row. This experiment by Slosson proved the deceiving power of smell. It is clear that cognitive factors related to learning and experience help guide the acuity of human olfactory perception (Gottfried and Wu, 2009).

Even the data from an expert research firm cannot promise success in sales

I have had personal experience with this when I was working for an ice cream development team. The problem was that I worked on a summer product development project during the winter and on a winter product development project in the summer. In a sense, it is not easy to develop a cold product for the mid-summer, while it is snowing outside. No matter how much we tried to imagine a hot, summer environment in the warm lab, soft products always taste better to the product developers during the cold season. However hard one imagines the environments of summer or winter, when the body feels summer the appetite prefers summer foods, and when the body feels winter the appetite prefers winter foods.

Smokers generally say that the smoke flavor feels different according to their body conditions. Even a non-smoker has probably experienced that a food tastes bad when he or she catches a cold or is queasy. This is because the sense of smell is affected by the body's condition. The olfactory perception of extremely sensitive people is affected by the weather, temperature, and humidity. Because the olfactory threshold changes, depending on the conditions of their bodies or their surrounding environments, even the same odor can be recognized differently.

Enormous amounts of money are expended on new product development projects, with teams making tremendous efforts to decrease individual deviations and to produce unbiased results to increase their success rates. Thus, food companies outsource huge sensory evaluation projects to expert research firms. The sensory experts try to minimize test errors, in other words, minimize the influence of various noise factors that might affect the sensory results. Since the perceived flavor changes according to the order of the test samples, they control the comparison group, centralization tendency error, time dependency, and location influence through various experimental designs and statistical analysis methods. Psychological deviation in particular is treated as a significant factor. They also carefully control the expectation error (if the goal is cost reduction), stimulation error (if it is held in a luxurious container), logic error (if the color is different), and halo effect (if it tastes good, the rest of the examining factors are also evaluated as good). Therefore, the sensory evaluations are conducted individually in independent booths.

A sensory lab is carefully designed to be equipped with proper lighting, ventilation, temperature, and humidity, as is the quality of the interior finishing material. The lab is even controlled to have the containers to provide the correct sample, sample size, serving temperature, and palate cleansing water. The test samples are not given serial numbers such as 1, 2, 3 or A, B, C. Since even the numbers and characters might affect the evaluation, random numbers and letters, for example, 683, 194, 327 or CBA, BCA, BAC, CAB, are assigned to eliminate the possibility of unconsciously favoring A, 1, or any other preferred numbers. After all this, data are analyzed using specialized statistical software. This is why only specialists can conduct sensory evaluation projects. However, product developers should not be too confident that only good results will came back from such specialized research firms. The success rate of a new product is generally not higher than 5%. A specialized research firm increases the success rate of a new product, but many other variables still exist.

In reality, the test itself is not natural, and there is no way that taking the test will not affect our sensing ability. Our senses are also affected by type of language that is used. The thinking process is the same process as using a language. However, sometimes language might not only be unhelpful for memory (decision-making) but can also actually make the situation worse. When a person who has a higher intellectual ability than language ability uses words to describe something, the "language barrier" phenomenon occurs, which interferes with the cognitive ability. This is already a well-known phenomenon in the field of cognitive science. It appears in various forms of stimulation such as visual media, color, and music. For example, people have the excellent ability to recognize the faces of other people. However, most people are very unskilled at describing a face with words. If we are forced to describe the face of an unknown person with words, we have a tendency to lose the important visual information of the face that was used in the recognition process and we cannot remember the details of the unknown person's face. A description that disappears in this way is generally information that is harder to express in words. This phenomenon also happens with smell or taste. Because it is really difficult to describe smell and taste in words, the effort required to describe them will lead to the loss of non-verbal sensory information. A person who frequently drinks wine can easily identify the wine he drank five minutes ago. However, if they must describe the taste with words after tasting it, the person has difficulty in selecting the same wine correctly just after the description test, because of the memory loss of the taste of the previous wine. This is because verbal description over-writes all the necessary information normally used to distinguish the taste of the wine. To overcome such language barriers, high-level training is required. Just like sommeliers, experts who are trained in description analysis are very good at duplicating the taste description. This is because continuous training and solid knowledge allow them to avoid the barriers that are presented. When they are asked to pick the wine they just tasted using multiple choice, they accurately select the right one. So when the trained experts are asked to describe the wine they just tasted verbally, their distinguishing ability is not diminished at all. This is because they always use their descriptive language containing the fundamental information that is designed to analyze the taste and flavor components from their combinations.

Sensorial perception is an illusion

The superiority of humans comes from the fact that we move less. If there were ten sensory neurons, there would be 100 000 interneurons, but only one motor neuron. The first ten cells feel, the other 100 000 recall memory, and deliberate to take the minimum but the most appropriate action. Reflexive reaction to a sense is not appropriate for the sensorial perception of a higher animal. However, we cannot completely trust our senses.

It is hard to understand why Koreans love spicy chili peppers more than any other country. Their chili pepper consumption rose 40% from 5.2 g per person a day in 1998 to 7.2 g in 2005. The annual consumption per person is 2.6 kg, a world record. Why do they love hot chili pepper even though it is so spicy?

In 1997, Professor David Julius's team at the University of California, San Francisco, solved this secret (Caterina *et al.*, 1997). TRPV1, one of several heat sensors in our bodies, detects high temperatures over 43 °C in order to prevent burns. When we eat hot chili, capsaicin in the chili pepper binds with TRPV1 and sends heat and pain signals to the cerebrum. TRPV1 also senses the pungent chemical of wasabi and horseradish, that is, allyl isothiocyanates. There is no actual temperature increase from these chemicals, TRPV1 just sends the heat signals. The brain, thinking that we are exposed to a high temperature, reacts to lower the heat and makes the body sweat. From the brain's perspective, it is completely fooled, but we do feel heat after eating chili peppers and we use the descriptive word hot as a synonym for spicy. Endorphin is released after this heat and pain perception. This is a chemical substance released naturally from the brain when our body feels pain, which has a pain-relieving effect 100 times stronger than narcotics. This is why people are addicted to spicy foods. The more capsaicin is consumed, the more stimulation to TRPV1 occurs and the brain releases more endorphin, and overall the level of pleasure to the spiciness increases.

It has been explained in previous chapters that capsaicin is originally a chemical weapon made by the chili pepper to protect itself from animals. Every vertebrate has the TRPV1 heat sensing channel, and feels pain from capsaicin. However, birds are calm after eating chili peppers, as the TRPV1 structure of birds is different to that of mammals. As a result, TRPV1 in birds detects heat, but does not react to capsaicin. They cannot feel the spiciness (pain) from chili. When a chili pepper wants to spread its seeds, it must seek the help of animals. However, unlike birds, mammals have teeth that chew the seeds so the seeds cannot sprout. In addition, to spread the seeds further, flying birds are better partners than walking mammals. Thus, the chili drives away unwelcome mammals using capsaicin.

Another interesting temperature receptor is TRPM8, which is a cold sensor that detects temperatures below 25 °C. Its activity increases as the temperature decreases. Coincidentally, menthol, the main component of peppermint, stimulates TRPM8. Many people used to think that the cooling sense of menthols was due to the flavor, but after the discovery of TRPM8, it was revealed that it is really a result of its effect on the sense of temperature. Also, people feel temperatures lower than 15 °C as pain. This is because of the function of TRPA1, which also reacts to the spicy substance allicin in garlic or isothiocyanates in wasabi. It was found that TRPA1 is distributed in the same nerve cells as TRPV1, which reacts to the capsaicin of chili. These sensors are different, but they are both the reasons why we feel "spiciness."

If a substance that binds to the temperature receptor TRPV4 were to be found, it might be the beginning of a new trend in the food industry. TRPV4 reacts with relatively high molecular weight chemicals, which are neither volatile nor fragrant. It is a temperature receptor that feels atmospheric temperatures, which can be applied to various functions. For example, it can give a slight warm feeling to cold ice creams, while giving a cooling sense to hot soups. The substances and temperature receptors that have been discovered so far only react to either extremely high or very low temperatures: something being either too hot or too cold is painful. Thus, if a substance reacting with TRPV4 is neither cold nor hot it could be modified to give an appropriately mild sense of temperature, or, for example, when a soup is too hot or ice cream is too cold.

Taste is a form of synesthesia, in other words, an integration of multiple senses, which also has a very flexible plasticity. Neuroplasticity gives hope to a person whose brain is damaged and synesthesia can make a person become a literary or an artistic genius through their extraordinary sensory ability. Our sensory ability depends on how much we use it and train it.

Taste and aroma do not exist

Research work into how cacao fruits produce their irresistible flavor and taste has been published. Professor Peter Schieberle's team at the Technical University of Munich found that out of hundreds of components in the cacao bean, only 24 are related to the taste of chocolate (Frauendorfer and Schieberle, 2006). These 24 individual components did not smell like chocolate but were more similar to other odors such as potato chips, cooked cabbage, peaches, undercooked beef fat, cucumber, honey, soil, and human sweat. However, a combination of the 24 chemicals gives a taste exactly like chocolate with no statistical difference from the original chocolate flavor. Of course, chocolate flavor is not made in this way, as real cocoa powder is not expensive, so large amounts can be used to make chocolate products. However, it is important to know that taste can be created through combinations of relatively small numbers of chemical substances.

Using three G-receptors, we can sense a very narrow spectrum of light so that we feel color. Each animal senses different wavelengths of light, so they feel color differently. We cannot say that our vision is correct and theirs is wrong, and it is the same with taste. It is not true to say that sugar is sweet, but it is one of the G-receptors responsible for our taste that binds with the sugar molecule and sends an electrical signal to the brain. Similarly, the taste of umami is just a phenomenon where the G-receptor is switched on when it binds with glutamic acid and sends electrical signals to the brain. There are no substances that actually possess sweetness or umami. It is simply that our brain feels that some substances have sweetness, umami, sourness, saltiness, or bitterness, when the appropriate G-receptors deliver signals to the brain because our body requires these substances as essential nutrients. Taste does not exist in the chemical substance, but it is how it is perceived by our brains.

A good product image makes it taste better

The phenomenon of our sense of taste and smell is an integration of a mixture of a wide range of senses and is the result of the brain having the plasticity to be able to transform to a considerable extent, depending on the situation. Therefore, the expectation of its absolute consistency is just a dream. Evolution pursues a decent and appropriate level of development, rather than perfection, because perfection costs too much. However, such inaccuracies and illusions can be utilized for a good purpose as the senses depend on how much they are used and as the brain can easily control the mouth and nose, there is no such thing as an absolute taste.

To illustrate this, consider the story of two professional baseball teams in the Korean League, the Tigers and the Giants, who formed an interesting rivalry in the 1980s. The teams were owned by the two biggest food companies in Korea, Haitai and Lotte, respectively. In the Tigers' hometown, the Haitai Company products sold well. In the Giants' hometown, the Lotte Company snacks sold better than the Haitai products. When an expert research institute carried out a survey to investigate whether regional preferences were considered in the decision-making of snack purchases, every respondent said that this had no effect. Haitai hometown residents said that they eat Haitai snacks simply because they are tastier than Lotte snacks. However, Lotte hometown residents said that Lotte snacks tasted better than Haitai snacks. The respondents answered that regional boundaries did not affect their choice of snack purchases and that the only reason was due to a taste preference. However, it is unlikely that they would prefer the competing company's products. Perception controls feelings, and feelings control taste.

A negative image can make a product taste bad and reduces the pleasure. Being happy when consuming food is good for one's health. There is too much coverage in the media related to specific issues with various foods. Confrontational comments or sensitive scenes are highlighted but for very limited examples, which do not give the whole story, and decrease the viewer's appetite. The media downgrade cheap foods by saying that they are fakes and contain low-quality ingredients and expensive foods are also criticized for the high price increases compared with the cost of the raw materials. They claim that safe additives in processed foods pose a toxic threat and that home-cooked foods have unsanitary ingredients. It is because of this type of information that our appetites are diminished and we must eat without joy and live in a time when it is the defective information that harms our healthy lives, not the defective foods.

References

Brochet, B. (2001) Chemical object representation in the field of consciousness. Application presented for the grand prix of the Académie Amorim following work carried out towards a doctorate from the Faculty of Oenology, General Oenology Laboratory, 351 Cours de la Libération, 33405 Talence Cedex. stevereads.com/papers_to_read/brochet_wine_experiment.pdf (accessed 17 July 2014).

Caterina, M.J., Schumacher, M.A., Tominaga, M., Rosen, T.A., Levine, J.D., Julius, D. (1997) The capsaicin receptor: a heat-activated ion channel in the pain pathway. *Nature.* **389**(6653): 816–824.

Dalton, P. (2012) There's something in the air: Effects of beliefs and expectations on response to environmental odors. In: *Olfactory Cognition: Advances in Consciousness Research* (eds. G.M. Zucco, R.S. Herz, B. Schaal), John Benjamins, Amsterdam, Holland, pp. 23–38.

Escoffier, A., Gilbert, P., Fétu, E., Suzanne, A., Reboul, B., Dietrich, C., Caillat, A. (1903) *Le Guide Culinaire, Aide-mémoire de cuisine pratique.* Émile Colin, Imprimerie de Lagny, Paris.

Frauendorfer, F., Schieberle, P. (2006) Identification of the key aroma compounds in cocoa powder based on molecular sensory correlations. *J. Agric. Food Chem.* **54**(15): 5521–5529.

Gottfried, J.A., Wu, K.N. (2009) Perceptual and neural pliability of odor objects. *Ann. N. Y. Acad. Sci.* **1170**: 324–332.

Slosson, E.E. (1988) A lecture experiment in hallucinations. *Psychol. Rev.* **6**: 407–408.

Yastrebtsev, V. (1908) On N.A. Rimsky-Korsakov's color–sound contemplation. *Russkaya Muzykalnaya Gazeta*, 1908, No. 39–40, pp. 842–845 (in Russian).

9 The Future of Taste and Aroma

Raw ingredient resources gradually become simplified and their original aromas disappear

Rice stalks lower their heads when they are mature and corn kernels remain on the shoots even when they are ripe. This may not seem strange, but, in reality, these types of rice and corn should not survive in nature. Normally, when they mature, seeds should fall down to the ground in order to germinate. However, rice and corn are mutants, and they have been modified to keep their seeds attached for the purpose of convenient and efficient harvesting. Humans have continuously selected and bred such mutants, through breeding technology, in order for these phenomena to occur. These mutant seeds have been spread intentionally, which means that the plants have become artificial species not found in nature, having been bred to keep their seeds intact. By nurturing these cultivars, the most preferred seeds are produced.

At the grocery store, most tomatoes boast a perfect red color and consumers definitely prefer these red tomatoes. However, they have neither a sweet nor a rich flavor because they are a mutant cultivar modified for uniform ripening, where the secretion of a protein (GLK2), which accelerates photosynthesis, is blocked. Because of the resulting slower photosynthesis, the tomatoes have lower sugar and nutrient contents, and a weak flavor. Such uniform ripening cultivars were discovered by tomato harvesters by accident in the 1930s. Since this red tomato color was preferred because of easy handling and long shelf life, the regular colored tomatoes disappeared from the

How Flavor Works: The Science of Taste and Aroma, First Edition.
Nak-Eon Choi and Jung H. Han.
© 2015 John Wiley & Sons, Ltd. Published 2015 by John Wiley & Sons, Ltd.

market. The latter are now only found in small vegetable gardens or at farmer's markets.

Today most milk comes from the breed of black spotted dairy cows called Holsteins. These cows are all fed with feedstuff manufactured by huge companies. Thus, cheese, as well as all other dairy products, has become more and more uniform. Chickens are a simplified breed, and are susceptible to massacre by avian influenza, something wild birds are not affected by at all. Eggs also come from a special breed of hens, for which egg production has increased threefold relative to wild hens. Cattle farmers say that a maximum of 58 L of milk can be obtained in one day from a dairy cow if the best quality feedstuffs are supplied. The goat and sheep breeds raised on dairy farms produce about 4 L of milk each day.

Thus, the business of agriculture today is controlled by productivity. In other words, cost determines everything. Only 25 types of grain out of 10 million crop species are utilized for human nutrition, and this trend is getting even worse. Continuous price and cost competition is drastically simplifying food ingredient resources and leaving only the most productive or high selling items on the market. Also, since the cultivation methods for these grains only emphasize their efficiency, their characteristic flavors have become weak and simple. Therefore, another aspect of human creativity (i.e., the technique of flavoring) is gradually becoming more important.

More scientific technologies will be incorporated into the culinary arts

Harold McGee, the author of *On Food and Cooking: The Science and Lore of the Kitchen*, says that "Cooking is applied chemistry." Thanks to him, cooking has entered the ivory tower and become an area of science, with the title of culinary science, and he has been applauded for enabling home and restaurant kitchens to practice this science. He studied literature, astronomy, and physics at Caltech and Yale, but after becoming interested in food science by chance, he spent the rest of his life grafting together science and cooking. Calling himself the "kitchen chemist," he has searched for the "why and how?" of everything related to cooking from the point of view of a chemist who has based his beliefs on "food, along with everything in the world, is a mixture of chemicals, and what we wish to influence in the kitchen such as taste, odor, texture, color, and nutrition are all of the properties of the chemical substances" (McGee, 1984).

The Research Chefs Association (RCA) registered a new terminology, Culinology®, which is the blending of culinary arts with the science of food, and they direct a recognized culinology degree program. RCA also manages professional certifications, which are Certified Research Chef (CRC) and Certified Culinary Scientist (CCS). These culinology education programs and certifications provide evidence for the current trend in the blending of culinary arts and food science. As more advances in food science and technological knowledge are made, they will become incorporated into the culinary art experiences.

It is a well-known fact that all objects around us consist of various molecules. Recently, food scientists also introduced the molecular concept. The aims of molecular cooking and molecular gastronomy are to increase the safety, taste, and visual effects of a product through physics or chemistry. The origins of molecular gastronomy were in 1988 with the, late, Oxford physicist Nicholas Kurti and the French INRA chemist, Hervé This. They conducted research and developed projects to design new equipment and cooking methods to find the hidden secrets of traditional European foods and methods of cooking. Their work could be called an invention and it was a conceptual shift towards breaking down our basic stereotypes of foods through physical and chemical experiments. However, it is not a complete demolition of the structure of the present system of food and cooking. It is merely a reinterpreting through a new paradigm referred to as science for the purpose of finding the essence of taste.

The fundamental premise of molecular gastronomy is to change the physical form of the ingredients quite significantly, while maintaining the taste and odor. For example, a striking display may be provided by a beautiful piece of art, but this could be fooling the viewer as to its true value, and in the same way, taste can sometimes betray the expectations that are anticipated from the appearance of a food. An apple shaped ice cream could taste like kimchi and what looks like caviar could taste like mangos. Someone might think and say, "You are playing with foods to fool people." However, both the credibility and the future of the food industry lie in this statement, because the taste and flavor of foods are simply what our brains perceive them to be. Molecular gastronomy actually understands the answer to the question of "how taste and flavor work." Foods are simple and the taste and flavor are just the outward appearance of the chemical ingredients. The essence (molecules of the ingredients) does not change just because the presentation is changed. However, processed or cooked foods do not have the same tastes and flavors as the natural raw ingredients as most tastes are flavors created through heating. Another example is the action of dipping sashimi in soy sauce, which makes the final taste and flavor of the sashimi seem completely unnatural. The sashimi is chosen carefully from the fish, with the relevant part being cut into the appropriate sized pieces, and the pieces are then dipped into the most artificial of flavoring materials, fermented soy sauce. Thus compositing the tasting and flavoring molecules for food safely is the most important scientific fundamental of molecular gastronomy.

What is the difference between cooking and the processing of foods?

If someone says that the taste of a processed food is good because it contains additives, most people would respond that this statement is wrong. However, when freshly cooked food tastes worse than processed food, this means that the chef did not do his or her best work during the cooking. In any situation, food is the most delicious when it is freshly cooked as its taste and quality will decrease with time. The loss of flavor is the main cause of the deterioration in quality, and preference then decreases. However, sometimes taste improves when the food is aged, as in the case of fermented foods, because the stimulating, undesirable odors produced by various enzymatic reactions disappear during the aging process. On the other hand, processed foods are not designed to be consumed straight after production. They are foods that are expected to be consumed after distribution and over a certain period of time. Thus, it is not easy for processed foods to compete with freshly cooked foods. It is very important for a cook to bring out the characteristic flavors of each ingredient when they prepare a freshly cooked meal. Every ingredient changes its unique properties slightly over the season, but the cook should be able to select any seasonally fresh ingredients to produce the same food. Cooking allows for such variations, but in processed foods, everything must be identical and the taste must be constant for any given season. This is one of the fundamental principles of quality control, and is why many sciences have combined to solve this issue of variation of processed food products. It is easy to criticize processed foods in various ways, but there are no processed foods on the market that are actually easy to produce.

Until recently, cooking and science (especially chemistry) had not been associated with each other. Food prepared on a small scale, such as at home and in restaurants, following traditional recipes, had not received any scientific attention. Practical knowledge and recipes from thousands of years of experiences were enough to create some of the best foods required for running a small business or for feeding families with excellent food. However, food science has significantly improved the level of public health as well as the quality and market size of the food industry by turning it into a consumer packaged goods (CPG) business. All foods mainly consist of water, proteins, carbohydrates, and fat. Thus a simple understanding of the physical chemistry characteristics of these molecules will lead to a better understanding of food, with fewer errors and increased success rates, and further applications.

Eggs harden when they are heated and although this might not seem unusual, it is actually a very odd phenomenon. For most chemical compounds, when the temperature goes up, they become less viscous, they do not solidify. Hard-boiling an egg, kneading flour dough until it becomes elastic, whipping egg whites to make foams, and boiling soymilk to coagulate tofu are all examples of the same phenomena: the denaturation of the folded natural proteins, so that they unfold and network. The proteins that form the building blocks of living organisms are folded compactly in nature, exposing water-friendly amino acids to the surrounding aqueous phase and hiding the oil-friendly amino acids inside. When they are heated, the kinetic energy of the folded proteins, supplied by the heat energy, is greater than the binding energy, so they unfold their structures into random long coils exposing the lipophilic parts. These unfolded proteins mix well with other proteins or lipophilic materials in the aqueous phase and solidify to create a gel structure or to increase the viscosity. Once this aspect of protein denaturation is clearly understood, it is easy to comprehend several other physical chemistry phenomena, such as phase change, emulsification, and soft–solid rheology. Unfortunately, in reality, most people try to learn the physical chemistry of foods from experience (know how) rather than from the principles (know why) even though a simple understanding of the chemical properties and the rules relating to chemical bonds between carbon, hydrogen, oxygen, and nitrogen can explain most physico-chemical phenomena in food products.

Raising the level of understanding of the structural formula of molecules is like learning how to read a map. A person who masters map reading can find any destination with less trial and error than a person who does not have this skill. In the same way, a food product is a mixture of various chemical substances and the only shortcut to understanding the properties of the product is to develop the ability to assess the physical chemistry characteristics from their chemical structures and their interactions. This sounds very difficult, but there are not that many types of molecules in foods, so it is very helpful to know the chemistry of the food ingredients that are used the most. Food chemistry and the physical properties of food are very useful and relatively easy to learn for someone who has completed lower-level chemistry courses in college.

Aroma-releasing television or movies

Vision and sound are digitized in consumer electronic products, as music is a series of digitized sound waves and color on digital display screens is a mixture of three light waves (RGB). Recently, 3D displays have become popular, but attempts over a long period of time to produce 4D TV and movies, consisting of a 3D display and aroma release, have proved unsuccessful. The main reason is the complexity of the ingredients. In the case of color, red, blue, and green can be used to create any color on a monitor, and for sound, just changing the wavelength can produce a different sound. If flavors could be as simplified as this, it would be possible to create revolutionary applications and to produce any flavor very precisely.

Up to this point, expressing a flavor, has, at most, been a matter of quoting the flavor of a familiar ingredient. For example, "it tastes like garlic with some olive oil" is a very straightforward description. Thus, is there a way of dissecting the garlic flavor into describable odors? Is there any way to express the flavor with categorized odors, in the same way that green is decomposed into a little yellow mixed with blue? If this decomposing analysis could be achieved, then accurate expression of a flavor would be possible, and flavor training and odor combination would also be very easy as well.

Among the attempts to classify flavors, the one with the most potential was the odor system of Ernest Crocker and Lloyd Henderson. In 1927, they chose four basic facets of odors: fragrant, acis, burnt, and caprylic. Next, the presence of each of the facets within an odor was judged on a 0 to 8 point scale that expressed the general intensity of all four facets. They organized the flavors as reference standards in order to evaluate all other smells. For example, the rose received a 6 for fragrant, a 4 for acis, 2 for burnt, and 3 for caprylic. These four numbers (6423) are an identification number used to distinguish the rose odor. In the same way, vinegar is 3803 and freshly brewed coffee is 7683. This is similar to using numbered colors so that graphic designers and workers in the print industry can accurately communicate with each other. This odor system was quickly applied commercially, and soon liquor manufacturers, soap manufacturers, the US Army, and even the US Department of Agriculture used a standard set based on the Crocker–Henderson classification. However, it did not last long. This classification can apparently discriminate 6561 (=9^4) different odors, and some years later, Crocker rounded this up to 10 000, the number everyone has been citing ever since. However, in 1949 researchers at Bucknell University dealt a blow to this classification system (Gilbert, 2008). They showed that untrained people could not actually classify 32 reference samples using the main odors or arrange eight samples of one main odor according to their strength. User enthusiasm faded and the odor system also disappeared. There have been a couple more attempts to classify odors, but today, no more efforts are being directed towards creating a comprehensive classification system. Simplification of raw materials would be possible if the classification of odors was feasible, but this is still impossible due to the lack of information on the analysis of the mechanisms of the olfactory sense. What we now know is that there are 388 types of olfactory receptors with no common patterns, meaning that an odor is determined by a combination array of 388 on/off switches (=2^{388}).

Flavorists cannot totally imitate the complete flavor of a targeted food product even with 2000 fragrant chemicals. The flavor of watermelon, melon, and persimmon cannot even be closely imitated. Therefore, it is impossible to create a watermelon scent when a watermelon appears on TV. Even if the 388 basic odorous chemicals are identified and recipes for all of nature's odors developed, there would be the problem of where to store the 388 substances in the TV set box and how to spray them. Furthermore, there is the more difficult problem that the sensitivity differences between the concentrations of odorous substances in air can be up to a 1 million fold. Maybe this issue could be resolved if the odors were appropriately diluted and prepared beforehand, but the critical problem of the delivery and deodorization procedures still remains.

A gas molecule moves very fast. Table 9.1 shows the speed of some common gas molecules at room temperature. A molecule of ammonia moves at $600 \, \mathrm{m \, s^{-1}}$ at room temperature and under vacuum. This might sound unlikely when considering that an incredibly powerful hurricane moves at only $40 \, \mathrm{m \, s^{-1}}$, but it is correct. However, when ammonia is accidentally spilt in a corner of a science laboratory, it takes seconds, or up to a a minute, for a student in the opposite corner of the lab to smell it. In a traditional lab experiment for science classes, students do not immediately detect a spilt chemical substance even

Table 9.1   Speed of Gas Molecules at 25 °C and 1 atm

Molecules	Speed (m s^{-1})	Average Distance between Collisions (m)	Frequency of Collision in a Second
Hydrogen (H$_2$)	1770	1.24×10^{-7}	1.43×10^{10}
Oxygen (O$_2$)	444	7.16×10^{-8}	6.20×10^{9}
Diallyl sulfide	208	1.42×10^{-8}	1.50×10^{10}

if though it could be moving at $30 \, \mathrm{m \, s^{-1}}$. The reason this takes some time is because the molecules are surrounded by other gas and air molecules. Even in a dilute gas, there are a lot of other molecules in the air and since they constantly collide and change direction, the gas molecules can only move in a straight line over relatively short distances. As a result of these collisions, the ammonia particles in the chemical laboratory probably only move little by little. To cross the laboratory, even if each particle is moving very quickly before it collides, it can take a substantial amount of time to reach the olfactory sensory system of all of the students. This is why aroma-releasing TV would not be practical. It would be awkward if the smell of flowers were to be detected after the main character had already left the flower garden.

A problem that is much harder to solve than the release of an odor is deodorization. Video and audio are only electromagnetic radiation, but taste and smell actually involve molecules. Molecules do not disappear by themselves, but are slowly diluted over time. A much longer time is required for effective deodorization than the time required for release. If an odorous molecule is mixed with other earlier odors before it disappears, it becomes a completely different smell. However, if odor molecules moved faster without any collisions, we might not be able to detect their smells. Therefore, collisions and the changes in direction as a result of molecular diffusion are also important for the odor to be inhaled into the nasal cavities.

Is the taste of processed foods inferior to that of natural foods?

There is no technology in the world that can convert bad ingredients into good. However, it is not difficult to give an odor to an odorless ingredient, because an odor is simply due to the properties of a chemical. It is also easy to give color to a colorless ingredient. Again, this is because color is actually a property of a chemical substance. It also is easy to give taste to a tasteless ingredient. A delicious taste can be made from a combination of sweetness, saltiness, sourness, and umami. Thus, it is easy to provide the preferred taste, odor, and color to bland food products. Yet, there is no substance that can eliminate bitterness and nothing that can disguise a dirty odor as it is very difficult to mask such undesirable properties. Therefore, it would make sense if producers used expensive ingredients to create good tastes, and if consumers would be willing to pay the high prices. However, this does not work in practice. Consumers wish to buy cheap products made with expensive ingredients, while producers wish to sell expensive products made with cheap ingredients. Therefore, a reasonable compromise is to create good tastes using regular ingredients and to sell them at an appropriate price.

Even so, there are many people who disagree strongly with this statement. They are known as health campaigners, who say that many processed foods are fake foods. For example, they assert that imitation crabsticks are a fake food because they are not made from crabmeat but from white fish flesh and starch. Everyone knows that crabsticks are not actually crabmeat, it even says so on the packaging. Someone who expects to eat real crabmeat at the price paid for crabsticks is being naive. However, there is no difference between crabsticks and crabmeat in terms of nutritional value and safety, the only differences are in their prices, ingredient costs, and taste. If there were no noticeable difference between these three things, then the crabsticks would actually be worth eating. Although the distinctive flavor of crabmeat is due to a naturally occurring chemical substance, this does not mean that crabmeat is safer and healthier than the artificial crabsticks just because it is natural, as the chemical providing this flavor has no effect on health.

There is a very popular cold noodle dish in Korea made from buckwheat noodles in a cold beef broth. However, there was a TV program claiming that this well loved beef broth contained chemical flavor enhancers. If brisket and bones were used, the production costs for one serving size of cold noodles in a beef broth would be over $5.50. Taking into account the total cost of the noodles, seasonings, other ingredients, labor, the property lease, and initial equipment investment in addition to the $5.50 just for the broth , this product would be extremely expensive. On the other hand, if MSG is used, one serving size of beef broth can be made at a cost of 50 cents. It still tastes pretty good and the average person cannot detect any difference from the real beef broth. But how much would a person pay to eat this dish if the ingredients in broth alone cost over $5.50? Is it even worth paying that price for real beef broth? To answer this we have to ask two questions of ourselves. Do we enjoy the beef broth in the buckwheat noodle dish for its taste or for its nutritional value? If taste is the main reason, is it really worth paying $5.50 just for a thin broth, which is simply a particular mixture of sweetness, saltiness, sourness, and umami? This is why the same taste has been created with just sugar, salt, vinegar, some seasonings, and simple flavors, at a reasonable cost of much less than $5.50. If it is enjoyed for nutritional value, then beef broth is a very inefficient way of taking in the nutrients from meat. If this were the case, it would be better to eat the boiled meat, not the water the meat was boiled in.

The main premise of capitalism is that of added value. Only the best technologies can create products that are more expensive than cost to produce. Inefficient technologies may have very high production costs with low productivity. However, people are quick to believe that a cheap product is reasonably priced if it is produced in bulk. This is exploitation of labor by the manufacturer to keep prices low and it means that only the cheapest products are allowed to survive in the market. Nowadays, price comparisons have become very easy and consumers' expectations of product quality have risen. Markets should be based on the quality of the product. Yet this is a world in which resources are wasted as price competitions intensify, instead of differentiation being on product quality. A market based on price competition does not raise the satisfaction of either the producers or the consumers. Competition with respect to product quality may decrease the quantities produced and thus the market size. Nevertheless, the satisfaction of producers and consumers will increase through a market competing on quality. Reduction in the number of unnecessary products and improvement in quality would truly lead to an environment-friendly market strategy. However, the price margin can have several unforeseen ripple effects. A person selling expensive products with cheap production costs will pay more income taxes. All countries operate through the taxes they collect. So a citizen who pays more taxes contributes more to their society and country. It is considered that a person who has a high income through their technological expertize, and who pays more tax on this money, is more respected by society. In contrast, it is not good when a person who feeds his own greed through tax evasion to become richer and more powerful has a louder voice in society.

There are many stories about how hard it is to run a successful small business. Most local restaurant are small businesses, but there is another major issue when running a restaurant other than the economic difficulties. Many TV shows and other media outlets deal with food and restaurant reviews. They can have very positive influences on restaurants and food businesses. However, they do not always work constructively. For example, some of these reviews criticize the use of MSG in food. Even though people want stimulating tastes from restaurant meals and processed foods rather than plain tastes, certain review shows change their recommendation to a condemnation of a restaurant based solely on whether they use MSG or not. As indicated previously, the use of MSG in the food industry is harmless. Recommending or condemning a restaurant because of the use of MSG serves no purpose other than hammering a nail into the heart of an innocent restaurant owner.

Such shows have deviated from their original goal of protecting the public health by campaigning to reduce the use of harmful chemical seasonings. If a show was designed to satisfy this mission, they would try to find firm evidence of the harmful effects of MSG and request a ban on its use by the Food and Drug Administration. However, the sensationalist shows only judge and bully small independent business owners for their use of MSG despite the FDA's official statement that it is safe and harmless. They know that MSG is made through a fermentation process and is exactly the same as glutamic acid in water, which is the most common substance in our bodies. Also, the average daily consumption of MSG is 2 g, which is 1/1000 of the glutamic acid already present in our bodies (1.6 kg in an adult). Despite this, the reason these shows continue to irrationally criticize the use of MSG is to increase their popularity. Even though there is no scientific evidence of any harmful effects of MSG, viewers tend to believe the nonscientific opinions on the show more than actual scientific evidence. Just as people depend on the taste of umami, irrational health campaigners depend on public popularity.

Is it true that obsessions with flavors and seasonings have decreased?

There used to be a time when whole markets were dedicated to spices and seasonings. But has that era come to an end? Not really, as recently, for example, 1.08 kg of Italian white truffles were sold for $200 000. A truffle is only a fungus containing large amounts of steroid-based odorous components. Also, in 2003, to celebrate their 50th anniversary, Royal Salute Scotch whisky had a limited sales promotion of 255 bottles of "Royal Salute 50." Of these, 24 bottles were assigned to South Korea, which sold out in a short time despite the high price of $10 000 per 700 mL bottle. The difference between expensive hard liquor and regular alcoholic products is that hard liquor has almost twice the alcohol content, which is officially a carcinogen, but there is only a slight difference in the flavors. Is it really worth paying extra simply for more alcohol with special odorous chemicals? Why would people pay such a high price for temporary pleasure?

In late October 2012, an 8 year old British schoolboy, Charlie Naysmith, found a slippery yellow stone while walking on the beach. Charlie's parents found out that it was ambergris, not a rock, which is also called the jackpot of the sea. The 600 g yellow hard solid, which is merely excrement of a male sperm whale, turned out to be worth approximately £40 000 ($67 000).

Marbling is when there is white fat in red meat looks like a white flower has bloomed. Derived from the way the fat is thickly studded and with a marble pattern, today the official terminology to describe this is the marbling score, so a cut of beef is graded according to the marbling score. Obviously, other standards such as the meat and fat color are also considered, but the marbling score is the most important. In the 1980s, people did not pay attention to beef marbling, there was not even a general international standard for good marbling to indicate a good beef taste. However, beef with high marbling scores smells better when it is grilled and its texture is very soft and tender because there is more juice. For a steak, the inside temperature of a high marbling score beef cut does not go above 80 °C, but it gets very hot on the surface. Large amounts of roasting flavors are created when the temperature goes above 160 °C. In the case of pork, cooking bacon at this temperature could be called pan-frying in oil rather than cooking. Many people like golden brown and crispy bacon. However, what is the difference between meat that is grilled golden brown and crispy and meat that is fried? Although steak is said to be grilled, cooking steak with hot butter in a pan for a French style steak is actually more of a frying process.

Oil has a strong flavor-holding capacity. The flavors created during cooking are highly lipophilic and oil-soluble. If there is a high oil content in a food, that food has a lot of flavors and is more enjoyable. This is why meat with a good marbling score tastes better. Back when meat consumption was rare in most diets, meat was generally cooked in a broth or a stew, making meat with high a fat and marbling score less desirable. People only began to prefer a higher marbling score when the economy had become strong enough for them to have grilled meat in their diets daily. It is common knowledge that the true identity of marbling is oil and that there is an increased danger in consuming the oily flavors of grilled meat, but people still have not given up the aroma of high fat contents despite the health risks, such as high cholesterol and obesity.

Do technological developments of taste modifications induce obesity or become a key solution to the problem?

In the past, only the rich had the privilege of being obese. However, today, it is the poor people are more obese. According to the 2011 data announced by the Ministry of Health and Welfare of South Korea, the obesity rate for women in the low-income bracket was 28.9%, while in the high-income bracket it was 23.2%. Obesity is inversely related to household income, which has become a common phenomenon across the globe. Unfortunately, for the present generation, it is much cheaper and easier to get fast food than to prepare healthier food. Although there are many explanations and hypotheses, whatever the cause, low income, obesity, and fast food are related.

Unfortunately, there are many foolish explanations for some statistical analyses. For example, wild animals are diagnosed with cancer less than animals in a zoo, leading some to suggest that it is obvious that an artificial living environment is not as good as nature itself. However, the reason for a higher cancer rate in zoo animals is because they live longer than wild animals. Their average lifespan is up to three times that in the wild, and thus their extended longevity increases the possibility of diseases, including cancer. As another example, consider Japan, which is ranked second in the world for the highest processed food production; it is practically the first per capita. However, life expectancy is the longest in the world, and thus as life expectancy has increased, so has the number of cases of diseases such as cancer. Still many people believe that higher incidences of cancer and other diseases are because the consumption of processed foods has increased. But that is not true. That would be like arguing that disease rates increased because the number of hospitals increased, or the number of tooth cavities increased because toothpaste consumption increased.

Economic poverty is a factor in the deterioration of family relationships and stability, and poverty can cause obesity and a decline in academic grades. However, obesity is probably not a direct cause of failing grades. If there is a high correlation between the number of restaurants and the number of local residents in a district who are obese, then, did those people become obese because there are a lot of restaurants? Or did the number of restaurants increase because there are a lot of obese people who like to eat? The latter is probably true. Nevertheless, there are some theories that conclude that the problem of obesity is related to the number of restaurants. In fact obesity is directly affected by portion size, not by the number of restaurants. However, it is really difficult to reduce the portion sizes of meals. Many people binge eat when they are unhappy, and so being happier may decrease the obesity problem in this case. Thus modifying taste and flavor could either lead to technology being used to increase the happiness levels of the consumer or to technology that will leave consumers wanting more food. Clearly the most desirable technology for modifying taste and flavor would be one that makes consumers happier with smaller portion sizes, but there is little interest in this research area.

Technology of satiety: technology of cognitive science for taste and olfactory senses is the technology of the future

Most problems caused by food products are related to obesity from overeating. This problem may be solved by changing the DNA in our body to feel sated 30% earlier, or by making products that are more filling, making us eat 30% less. Since there is no gene therapy technology to change human DNA, we must research ways of increasing our feeling of satiety. To solve the obesity problem, we must focus on satiety more than on calorie intake, yet most research projects concentrate on calorie intake only.

Bananas have the most calories among the common fruits and are the only fruit that contains fat. So if calories are the problem, we should be eating other fruits, not bananas. However, many weight control programs recommend eating bananas because of their high level of satiety, which is actually very effective. People can see the effectiveness of fat consumption on weight loss through low or restricted carbohydrate diet programs, which prove that fat is important for increasing satiety and that there is no relationship between fat consumption (i.e., high calories) and obesity under a controlled fat intake diet. However, people still focus only on calorie and fat reduction, which is why the solution to the obesity problem seems so far away.

Throughout human history, people have experienced difficulties in procuring foods, and food security is still a big research topic. Animals eat as much food as possible and store the energy as body fat for a time when food is scarce. It would be a form of suicide for animals in nature to pass by a sweet fruit or fresh meat without eating them. Nowadays people live in a world where cheap, abundant, and tasty foods are available everywhere. In such an environment, it is hard to suppress the evolutionary desire to eat as much as possible. It is actually very close to impossible under normal circumstances.

We require a technology that enhances satiety. Flavoring ingredients only have a value for sensing, not for nutrition. If there were a technology to satisfy the senses using flavors with only a small amount of food, it would be very useful for solving the obesity problem. Current food products do not have a problem with securing nutrition, but they do have more important issues of how to effectively satisfy the senses. All of these conditions are the reasons for redirecting food research towards satiety. Satisfying the senses is the most valuable aspect of the food business and food consumption. However, amazingly, there are only a few research projects and researchers working on how to enhance satiety. Although there are examples related to the satiety control, they still need to be investigated for their long-term effects.

Satiety control

Eat strongly flavored foods: There is an investigation that has reported that eating strongly flavored foods is useful for eating less and losing weight. In March 2012, a research group in The Netherlands found that we could eat 5–10% less if the food we eat has stronger flavors (de Wijk *et al.*, 2012). Although the research team made the flavor intensity slightly higher than the control so that the participants could barely distinguish the difference, when the flavor was stronger the participants thought that the food had a lot more calories and consequently felt more satisfied. From this result alone, we should not conclude that increasing flavor intensity is essential for decreasing food consumption. However, it is certain that the lack of satisfaction and satiety leads to overeating. It has been known for a long time that when people continuously eat monotonous foods, they then look for strongly flavored foods. In 2004, according to Marcia Pelchat at Monell Chemical Senses Center (Pelchat *et al.*, 2004), it was specifically confirmed that continually eating simply flavored and monotonously flavored meals led to an increased desire for foods with rich tastes and flavors.

Emulsifiers or stabilizers can double the duration of satiety:
Dr Marciani's research team at Nottingham University revealed that using emulsifiers or stabilizers on regular food can double the duration time of digestion in the stomach and thus the feeling of satiety (Marciani *et al.*, 2009). Participants in the experiment remained satiated for 12 hours after drinking a sample based on olive oil, water, and/or stabilizer (hydrocolloid). This was given in the form of a milkshake to 11 volunteers and MRI scans were conducted hourly to check how much of the drink remained in the stomach. After one hour, the results showed that in people who drank the test sample in which the oil and water phases were not separated (i.e. with the stabilizer), there was twice the volume of the sample left in their stomachs than the group who drank the sample without the hydrocolloid stabilizer. The sample remains even longer if the oil and water are homogenized, which leads to more satiety and decreases the desire for food. This experiment carries many implications. It is a well-known fact that blended foods lead to prolonged satiety and are helpful in weight management. However, grinding food together takes away the joy of eating, but consuming appropriately emulsified food induces good satiety. This is the same principle that applies to eating a salad with an appropriate amount of dressing being effective in weight management.

Eating a variety of foods together: When eating several different types of food together we consume the same total calories no matter how they are eaten, although satiety lasts longer when those foods are absorbed slowly in the gastrointestinal track. Therefore, eating foods containing protein and fat with a large quantity of vegetables could retard the food absorption. However, some types of food, particularly carbohydrates, are absorbed quickly in the gastrointestinal track, making us hungry sooner.

Eating high moisture foods or foods with water: When eating carbohydrates in water or consuming foods with a high moisture content, the total amount of solid per volumetric consumption is decreased. Therefore, the total calories are reduced. Also, eating foods along with a large volume of water causes dilution of digestive enzymes and, therefore, the absorption of nutrients is interrupted.

Eating peels of fruits: Fruit peels contain a lot of fibers. Eating fruits with peel decreases the glycemic index, and retards insulin activity. Insulin quickly reduces the sugar level and leads to hunger. However, retarding the insulin activity through the fruit peels extends the satiety longer than eating fruits without peels.

Eating high protein foods: High protein meals stay in the stomach longer and delay digestion. Therefore, they maintain the fullness feeling of the stomach and extend satiety.

Eating soluble fibers: Water soluble dietary fiber increases the viscosity of the food mass in the intestine and lowers the digestion rate to keep the volume of food constant. Inclusion of dietary fiber in food decreases the feeling of emptiness in the stomach and thus extends satiety.

These are examples of very limited research results for enhancing satiety compared with the efforts made to make food tastier. Everyone is looking for a magic solution that can comfortably solve the problem of obesity instantly rather than concentrating on the extension of satiety. However, there are over 3000 causes of obesity. Efforts to solve obesity by finding the cause in the food components have been made over the last couple of decades but as yet no conclusions can be made for this approach.

The era of supernormal stimuli

If we knew the taste, smell, and preferred characteristics, we could make a replica that is more realistic than the original. However, if we did make something more realistic than the original, we would get something supernormal, not abnormal. This is an era of supernormal stimulus where people take abnormal for normal. Nikolaas Tinbergen, a Dutch ethologist and ornithologist, the 1973 Nobel Laureate in Physiology or Medicine, coined the expression "supernormal stimulus" after confirming that an imitation made by an experimenter can stimulate an animal instinct more than its normal stimuli. For example, cuckoos use the supernormal stimulus of other birds to hatch their eggs. They are famous in the animal world for parasitic baby care, or in other words "brood parasitism" (a bird putting its egg in another bird's nest to be nurtured and hatched by the other bird). The cuckoo puts its eggs in the nest of the host birds and allows them to raise their chicks instead of doing it themselves. The cuckoo's egg is slightly bigger and brighter than the eggs of the host birds, making them want to sit on the cuckoo's eggs more than their own eggs. There are an innumerable number of other such examples of supernormal stimulus.

Nowadays, we also live in a supernormal era. We also know how to satisfy our desire for taste, as we feel sweetness for carbohydrates and umami for proteins. This is an era where the secret of taste and flavor has been discovered and strengthened to give a much stronger satisfaction (supernormal stimulus). Society is not interested in why we like symmetrical and S-shaped bodies, but only concentrates on how to make this body shape. All women are becoming identical in their appearance and body shape, and food tastes are becoming uniform nationwide to a single taste that is statistically known as the best.

In the past, tasty food was food that was good for our health. Now, we ignore the nutritional values and depend on our senses too much. People used to cook foods to enjoy their tastes and flavors and, therefore, their health benefits. Today we cook foods to create good tastes and flavors, disregarding their nutritional values. It is an era when foolish expressions, such as "tasty food is not good for you," which has the same meaning as "if it tastes bad, it's good for you," are generally accepted. Yet this is truly an abnormal situation. It is still true that tasty food is good food. Problems related to food are caused by overeating of tasty food that is good for your health. Thus most problems are related to our extreme eating habits, not to the foods themselves. However, our desires have replaced our problems with a food quality issue. We live in an era with the highest anxiety about food products, even though we are eating the safest food products in the history of mankind.

References

de Wijk, R.A., Polet, I.A., Boek, W., Coenraad, S., Bult, J.H.F. (2012) Food aroma affects bite size. *Flavour.* 1: 3.

Gilbert, A. (2008) *What the Nose Knows: The Science of Scent in Everyday Life.* Crown Publishers, *New York.* ISBN-13: 978-1400082346, p. 21.

Marciani, L., Faulks, R., Wickham, M.S., Bush, D., Pick, B., Wright, J., Cox, E.F., Fillery-Travis, A., Gowland, P.A., Spiller, R.C. (2009) Effect of intragastric acid stability of fat emulsions on gastric emptying, plasma lipid profile and postprandial satiety. *Br. J. Nutr.* **101**(6): 919–928.

McGee, H. (1984) *On Food and Cooking: The Science and Core of the Kitchen.* Scribner, New York.

Pelchat, M.L., Johnson, A., Chan, R., Valdez, J., Ragland, J.D. (2004) Images of desire: Food-craving activation during fMRI. *NeuroImage,* **23**, 1486–1493.

10 Concluding Remarks

In conclusion, there are only five tastes: sweetness, saltiness, sourness, bitterness, and umami. The various tastes we actually perceive are in fact flavors. Thus, we can say that 90% of taste is flavor. However, less than 50% of tastes are perceived through the mouth and nose as sensing through the intestines is equally as important. The GI track has enough sensorial cells to be called a second brain. It rates the quality of the food we eat and senses not only salt, amino acids, and glucose, but also fatty acids, and sends signals to the sensory cortex in the brain. Other than nutritional content, it also detects the amount, osmotic pressure, temperature, shape, size, and texture of the food. The actual number of sensory receptors in the intestines is greater than the number of sensory receptors on the tongue. Thus it is through these senses that our brains know what we have eaten and they can then carry out a series of digestive processes efficiently, such as secreting digestive fluids. The sensorial ability of our GI tracks also makes us feel satiated. Because we only think that we taste through our mouth and are unaware that the GI track can detect the amount of fat, we can make foolish choices. Our body immediately recognizes low fat foods or even food products that have exactly the same taste as their full fat counterparts through the sensorial ability of our GI tracks.

However, even though many low-calorie, low-salt, and low-sugar products have been developed, most do not succeed in the market. The main reason they have such difficulty is because our bodies actually know how to enjoy calories, salt, sugar, and fat, and although appetite is important, the satisfaction of our internal organs is also significant. We may deceive our appetite by cutting back on fat and taking artificial sweeteners to decrease the total calorie intake, but due to the highly

How Flavor Works: The Science of Taste and Aroma, First Edition.
Nak-Eon Choi and Jung H. Han.
© 2015 John Wiley & Sons, Ltd. Published 2015 by John Wiley & Sons, Ltd.

sensitive nature of our body's senses, including the intestine receptors, it is very difficult to fool the body for very long with a low-fat and low-sugar diet.

When a woman becomes pregnant, her appetite drastically changes to satisfy the appropriate dietary needs for her and the fetus, but after giving birth, her diet returns to normal. There are many people who attempt to become vegetarians after hearing of the health benefits, but willpower alone is not enough to stick to a vegetarian diet for very long. Moreover, the body quickly recognizes the different diet, which does not satisfy the needs or expectations of the body. There is no technology that can deceive the peoples' senses over many years. If there were such a technology that could fool the senses of the human body, the solution to obesity would be at hand.

However, even with the five senses and the intestine receptors, all these sensing mechanisms only explain less than half of the cognitive reactions to tastes and flavors. The rest is due to the work of the brain. Readers of this book may have a vague knowledge of psychological effects on taste perception, but the author hopes that the book has helped awaken a greater interest in how much the brain (the mind) influences the final perception of tastes.

Recently, a book was published that shocked me greatly. It was *Why Humans Like Junk Food* (Witherly, 2007). Everyone says that junk food is bad. Yet there have been no studies of junk food that explain why these businesses are so successful under the face of such criticisms. There are only conventional and irrational criticisms that they have captured the consumers' appetites through the use of various additives. On Wikipedia, junk food is defined and explained as "a derisive slang term for food that is of little nutritional value and often high in fat, sugar, salt, and calories. It is widely believed that the term was coined by Michael Jacobson, director of the Center for Science in the Public Interest, in 1972. Junk foods typically contain high levels of calories from sugar or fat with little protein, vitamins or minerals. Foods commonly considered junk foods include salted snack foods, gum, candy, sweet desserts, fried fast food, and sugary carbonated beverages. Many foods such as hamburgers, pizza, and tacos can be considered either healthy or junk food depending on their ingredients and preparation methods. The more highly processed items usually fall under the junk food category."

However, this suggests that we are sensible and reasonable. If we were this sensible, we would know the exact reason why people like carbonated beverages, French fries, and hamburgers. Cola was introduced into the market over 100 years ago. If the cola business was successful because of its additives or ingredients, there should be no reason why others could not imitate its fantastic taste, since all the food additives and ingredients are clearly defined and labeled. In the same way, French fries are simply fried potato, and hamburgers do not actually use any preservatives or special additives either. If food additives created these tastes, this could be achieved without any greasy oil or salt, and thus easily avoiding the criticisms. So why are they still labeled as junk foods? Hamburgers and French fries do not contain added sugars, so why do people criticize them for inducing cavities in teeth when sugar is not even present? Why is no one interested in the fact that rice or juices creates more cavities than sugar? In reality, we are not sensible or reasonable judges.

Adults blindly tell their children to eat vegetables and fermented products. There are some children who readily eat them but others genuinely dislike them. However, have we ever seriously considered why children dislike them? Taste and smell are most sensitive among infants, more than at any other time their lives. During this period, there are taste buds all over the mouth and also taste receptors on the roof of the mouth, throat, and sides of the tongue. Because of this, infants can sense dilute, powdered milk to be much tastier than an adult would. The taste buds begin to disappear around the age of ten and the sensitivity of taste differs from person to person once they have grown up. As we have seen, a taste is detected by the taste buds. In a square centimeter, a less sensitive person has 100 taste buds, while an average person has 200, and a very sensitive person has 400. A sensitive person does not sense taste more accurately or precisely, but they simply become more sensitive to tastes, especially to the bitterness. About 25% of adults readily feel bitterness from typical foods. Therefore, the more sensitive children, with higher numbers of taste buds, will be many times more sensitive to bitterness. Vegetables contain high levels of toxins, which are bitter and approximately two-thirds of all amino acids taste bitter. In addition, taste umami from fermented products due to the glutamic acid and aspartic acid, but children might just taste the bitterness, making it hard for them to enjoy such fermented products. However, taste becomes less sensitive as children grow and they adapt to bitterness after becoming more familiar with bitter tasting foods. Thus, recklessly forcing a vegetable diet on children and saying that it is healthy without any consideration of the children's taste could be annoying.

This book deals only with the factors influencing taste and does not consider which ingredients and taste factors can be combined to provide a preferred taste. Taste is a fairly deep evolutionary development but it can also be utilized to run a successful business. If we know the particular formula for the taste factors of success, we can greatly reduce the trial and error involved, and even reduce the chances of being hurt by useless expectations.

Two previous books (*Ineligible Information Ruins Our Body* (Choi, 2012a) and *33 Secrets Of Foods* (Choi, 2012b), published in Korean) talk substantially about the safety of foods, while the present book and also subsequent books deal with the preference for foods. Thus the only big topic remaining is the nutritional value of foods. Although nutrition is the fundamental basis of foods and preference is a secondary variable for regulating the consumption of good nutrients, the reality is that the fundamental basis has disappeared and the secondary method has become the main goal. On the one hand, nutrition is a simple requirement but, on the other, it has received so much attention that it has become a problem. In short, although nutrients are chemical substances that our bodies need and are objects that are delivered through foods, we behave as if nutrients decide all the outcomes of our bodies, ignoring the body's adaptability, and we overreact to a slight temporary shortage in nutrition. Actually, no animals consciously try to eat an evenly balanced diet and there are no better living organisms in the world, other than humans, to be born with such excellent conditions to continue their survival.

There is simplicity deep within complexity. Simplicity is the pillar of our existence and it is the structure and harmony hidden in all creation. Many phenomena appear very complex but their principle is simple and often different phenomena actually have the same principles. The point is that we should understand the relationships between discovered knowledge. The secret to problem solving is not to simply understand a topic but also to understand the connections between topics. In order to become wiser, we must make efforts to overcome our dependency on perceptions and topics (i.e., facts). If we just understood that perceptions are illusions, we might become more rational about food and cut back on unnecessary expenses. We always surrender to perception, which is normal. However, striving to overcome the limits of our perceptions has always led to improvements. Let us attempt to overcome the limits of our perceptions more and more, because our perceptions are illusions in our brains.

We do not need to limit our joy of tasting delicious foods or smelling their rich aroma. Our brains can imagine the taste and aroma of certain foods that are displayed on a television monitor. All perception is an illusion that is created by our brains anyway and making the best use of those illusions is one of the best features of being human. Sometime in the future, maybe it will be possible for us to taste and smell in cyberspace. However, do not try to find a special feature or mystery from food components. The special feature is in the relationship between our bodies and the food components, not in the component chemicals.

Science is a simple exploration to explain Mother Nature, while art is the spring of the unleashed imagination. Food has strong cultural associations, positioned between science and art. This means that food must be evaluated both scientifically and artistically, as humans do not yet have the perfect knowledge to evaluate foods using only a scientific approach. However, many television programs change opinions through incorrect information on food science. These food programs should be evidence-based science programs, but they deal with the issues as if they were entertainment shows. If we accepted that all their points of view were true, there would be no healthy foods in the world and all of us would become patients who were sick due to eating unhealthy foods. Nowadays we are eating foods that are the safest and the healthiest they have ever been, which can consumed over long periods of time. If anyone wants to analyze foods scientifically, they must use scientific methods, but in most cases we just need to enjoy our food. Since all food ingredients and additives are substances that have been approved as being safe by the FDA, USDA (United States Department of Agriculture), and EPA (Environmental Protection Agency) in the United States, we just need to relax and enjoy all these perfectly safe foods without any worries.

References

Choi, N.E. (2012a) *Ineligible Information Ruins Our Body.* Geaho Publishing, South Korea.

Choi, N.E. (2012b) *33 Secrets Of Foods.* Kyunghyang Media, South Korea.

Witherly, S.A. (2007) *Why Humans Like Junk Food: The Inside Story on Why You Like Your Favorite Foods, the Cuisine Secrets of Top Chefs, and How to Improve Your Own Cooking Without a Recipe!* iUniverse, Inc. Bloomington, IN.

Index

Note: Illustrations are indicated by *italic page numbers*, Tables by **bold numbers**, and footnotes by suffix "n" (e.g. "124n" means "footnote on page "124")

7TM (transmembrane) receptors *see* G-protein coupled receptors

How Flavor Works: The Science of Taste and Aroma, First Edition.
Nak-Eon Choi and Jung H. Han.
© 2015 John Wiley & Sons, Ltd. Published 2015 by John Wiley & Sons, Ltd.

Food Science and Technology Books

WILEY Blackwell

GENERAL FOOD SCIENCE & TECHNOLOGY, ENGINEERING AND PROCESSING

Food Texture Design and Optimization	Dar	9780470672426
Nano- and Microencapsulation for Foods	Kwak	9781118292334
Extrusion Processing Technology: Food and Non-Food Biomaterials	Bouvier	9781444338119
Food Processing: Principles and Applications, 2nd Edition	Clark	9780470671146
The Extra-Virgin Olive Oil Handbook	Peri	9781118460450
Mathematical and Statistical Methods in Food Science and Technology	Granato	9781118433683
The Chemistry of Food	Velisek	9781118383841
Dates: Postharvest Science, Processing Technology and Health Benefits	Siddiq	9781118292372
Resistant Starch: Sources, Applications and Health Benefits	Shi	9780813809519
Statistical Methods for Food Science: Introductory 2nd Edition	Bower	9781118541647
Formulation Engineering of Foods	Norton	9780470672907
Practical Ethics for Food Professionals: Research, Education and the Workplace	Clark	9780470673430
Edible Oil Processing, 2nd Edition	Hamm	9781444336849
Bio-Nanotechnology: A Revolution in Food, Biomedical and Health Sciences	Bagchi	9780470670378
Dry Beans and Pulses : Production, Processing and Nutrition	Siddiq	9780813823874
Genetically Modified and non-Genetically Modified Food Supply Chains: Co-Existence and Traceability	Bertheau	9781444337785
Food Materials Science and Engineering	Bhandari	9781405199223
Handbook of Fruits and Fruit Processing, second edition	Sinha	9780813808949
Tropical and Subtropical Fruits: Postharvest Physiology, Processing and Packaging	Siddiq	9780813811420
Food Biochemistry and Food Processing, 2nd Edition	Simpson	9780813808741
Dense Phase Carbon Dioxide	Balaban	9780813806495
Nanotechnology Research Methods for Food and Bioproducts	Padua	9780813817316
Handbook of Food Process Design, 2 Volume Set	Ahmed	9781444330113
Ozone in Food Processing	O'Donnell	9781444334425
Food Oral Processing	Chen	9781444330120
Food Carbohydrate Chemistry	Wrolstad	9780813826653
Organic Production & Food Quality	Blair	9780813812175
Handbook of Vegetables and Vegetable Processing	Sinha	9780813815411

FUNCTIONAL FOODS, NUTRACEUTICALS & HEALTH

Antioxidants and Functional Components in Aquatic Foods	Kristinsson	9780813813677
Food Oligosaccharides: Production, Analysis and Bioactivity	Moreno-Fuentes	9781118426494
Novel Plant Bioresources: Applications in Food, Medicine and Cosmetics	Gurib-Fakim	9781118460610
Functional Foods and Dietary Supplements: Processing Effects and Health Benefits	Noomhorm	9781118227879
Food Allergen Testing: Molecular, Immunochemical and Chromatographic Techniques	Siragakis	9781118519202
Bioactive Compounds from Marine Foods: Plant and Animal Sources	Hernández-Ledesma	9781118412848
Bioactives in Fruit: Health Benefits and Functional Foods	Skinner	9780470674970
Marine Proteins and Peptides: Biological Activities and Applications	Kim	9781118375068
Dried Fruits: Phytochemicals and Health Effects	Alasalvar	9780813811734
Handbook of Plant Food Phytochemicals	Tiwari	9781444338102
Analysis of Antioxidant-Rich Phytochemicals	Xu	9780813823911
Phytonutrients	Salter	9781405131513
Coffee: Emerging Health Effects and Disease Prevention	Chu	9780470958780
Functional Foods, Nutraceuticals & Disease Prevention	Paliyath	9780813824536
Nondigestible Carbohydrates and Digestive Health	Paeschke	9780813817620
Bioactive Proteins and Peptides as Functional Foods and Nutraceuticals	Mine	9780813813110
Probiotics and Health Claims	Kneifel	9781405194914
Functional Food Product Development	Smith	9781405178761

INGREDIENTS

Fats in Food Technology, 2nd Edition	Rajah	9781405195423
Processing and Nutrition of Fats and Oils	Hernandez	9780813827674
Stevioside: Technology, Applications and Health	De	9781118350669
The Chemistry of Food Additives and Preservatives	Msagati	9781118274149
Sweeteners and Sugar Alternatives in Food Technology, 2nd Edition	O'Donnell	9780470659687
Hydrocolloids in Food Processing	Laaman	9780813820767
Natural Food Flavors and Colorants	Attokaran	9780813821108
Handbook of Vanilla Science and Technology	Havkin-Frenkel	9781405193252
Enzymes in Food Technology, 2nd edition	Whitehurst	9781405183666
Food Stabilisers, Thickeners and Gelling Agents	Imeson	9781405132671
Glucose Syrups - Technology and Applications	Hull	9781405175562
Dictionary of Flavors, 2nd edition	DeRovira	9780813821351

FOOD SAFETY, QUALITY AND MICROBIOLOGY

Practical Food Safety	Bhat	9781118474600
Food Chemical Hazard Detection	Wang	9781118488591
Food Safety for the 21st Century	Wallace	9781118897980
Guide to Foodborne Pathogens, 2nd Edition	Labbe	9780470671429
Improving Import Food Safety	Ellefson	9780813808772
Food Irradiation Research and Technology, 2nd Edition	Fan	9780813802091
Food Safety: The Science of Keeping Food Safe	Shaw	9781444337228

For further details and ordering information, please visit www.wiley.com/go/food

Food Science and Technology from Wiley Blackwell

Decontamination of Fresh and Minimally Processed Produce	Gomez-Lopez	9780813823843
Progress in Food Preservation	Bhat	9780470655856
Food Safety for the 21st Century: Managing HACCP and Food Safety throughout the Global Supply Chain	Wallace	9781405189118
The Microbiology of Safe Food, 2nd edition	Forsythe	9781405140058

SENSORY SCIENCE, CONSUMER RESEARCH & NEW PRODUCT DEVELOPMENT

Olive Oil Sensory Science	Monteleone	9781118332528
Quantitative Sensory Analysis: Psychophysics, Models and Intelligent Design	Lawless	9780470673461
Product Innovation Toolbox: A Field Guide to Consumer Understanding and Research	Beckley	9780813823973
Sensory and Consumer Research in Food Product Design and Dev, 2nd Ed	Moskowitz	9780813813660
Sensory Evaluation: A Practical Handbook	Kemp	9781405162104
Statistical Methods for Food Science	Bower	9781405167642
Concept Research in Food Product Design and Development	Moskowitz	9780813824246
Sensory and Consumer Research in Food Product Design and Development	Moskowitz	9780813816326

FOOD INDUSTRY SUSTAINABILITY & WASTE MANAGEMENT

Food and Agricultural Wastewater Utilization and Treatment, 2nd Edition	Liu	9781118353974
Sustainable Food Processing	Tiwari	9780470672235
Food and Industrial Bioproducts and Bioprocessing	Dunford	9780813821054
Handbook of Sustainability for the Food Sciences	Morawicki	9780813817354
Sustainability in the Food Industry	Baldwin	9780813808468
Lean Manufacturing in the Food Industry	Dudbridge	9780813810072

FOOD LAWS & REGULATIONS

Guide to US Food Laws and Regulations, 2nd Edition	Curtis	9781118227787
Food and Drink - Good Manufacturing Practice: A Guide to its Responsible Management (GMP6), 6th Edition	Manning	9781118318201
The BRC Global Standard for Food Safety: A Guide to a Successful Audit, 2nd Edition	Kill	9780470670651
Food Labeling Compliance Review, 4th edition	Summers	9780813821818

DAIRY FOODS

Lactic Acid Bacteria: Biodiversity and Taxonomy	Holzapfel	9781444333831
From Milk By-Products to Milk Ingredients: Upgrading the Cycle	de Boer	9780470672228
Milk and Dairy Products as Functional Foods	Kanekanian	9781444336832
Milk and Dairy Products in Human Nutrition: Production, Composition and Health	Park	9780470674185
Manufacturing Yogurt and Fermented Milks, 2nd Edition	Chandan	9781119967088
Sustainable Dairy Production	de Jong	9780470655849
Advances in Dairy Ingredients	Smithers	9780813823959
Membrane Processing: Dairy and Beverage Applications	Tamime	9781444333374
Analytical Methods for Food and Dairy Powders	Schuck	9780470655986
Dairy Ingredients for Food Processing	Chandan	9780813817460
Processed Cheeses and Analogues	Tamime	9781405186421
Technology of Cheesemaking, 2nd edition	Law	9781405182980

SEAFOOD, MEAT AND POULTRY

Seafood Processing: Technology, Quality and Safety	Boziaris	9781118346211
Should We Eat Meat? Evolution and Consequences of Modern Carnivory	Smil	9781118278727
Handbook of Meat, Poultry and Seafood Quality, second edition	Nollet	9780470958322
The Seafood Industry: Species, Products, Processing, and Safety, 2nd Edition	Granata	9780813802589
Organic Meat Production and Processing	Ricke	9780813821269
Handbook of Seafood Quality, Safety and Health Effects	Alasalvar	9781405180702

BAKERY & CEREALS

Oats Nutrition and Technology	Chu	9781118354117
Cereals and Pulses: Nutraceutical Properties and Health Benefits	Yu	9780813818399
Whole Grains and Health	Marquart	9780813807775
Gluten-Free Food Science and Technology	Gallagher	9781405159159
Baked Products - Science,Technology and Practice	Cauvain	9781405127028

BEVERAGES & FERMENTED FOODS/BEVERAGES

Encyclopedia of Brewing	Boulton	9781405167444
Sweet, Reinforced and Fortified Wines: Grape Biochemistry, Technology and Vinification	Mencarelli	9780470672242
Technology of Bottled Water, 3rd edition	Dege	9781405199322
Wine Flavour Chemistry, 2nd edition	Bakker	9781444330427
Wine Quality: Tasting and Selection	Grainger	9781405113663

PACKAGING

Handbook of Paper and Paperboard Packaging Technology, 2nd Edition	Kirwan	9780470670668
Food and Beverage Packaging Technology, 2nd edition	Coles	9780813812748
Food and Package Engineering	Morris	9780813814797
Modified Atmosphere Packaging for Fresh-Cut Fruits and Vegetables	Brody	9780813812748

For further details and ordering information, please visit www.wiley.com/go/food